ELECTRICIAN'S EXAM STUDY GUIDE

About the Authors

B. D. COFFIN is a Master Electrician and Fire Alarm Installation Trainer with over 30 years of experience in the residential, commercial, and industrial markets.

K. J. KELLER is the CAD Designer, Safety & Compliance Director, and Project Coordinator for a leading New England electrical contractor. She has over 30 years of experience in residential and commercial construction.

ELECTRICIAN'S EXAM STUDY GUIDE

B. D. Coffin and K. J. Keller

New York   Chicago   San Francisco   Lisbon   London   Madrid
Mexico City   Milan   New Delhi   San Juan   Seoul
Singapore   Sydney   Toronto

The McGraw·Hill Companies

Library of Congress Cataloging-in-Publication Data

Coffin, B. D. (Brian D.)
 Electrician's exam study guide / B. D. Coffin and K. J. Keller.
  p. cm.
 ISBN-13: 978-0-07-148930-0
 ISBN-10: 0-07-148930-4 (alk. paper)
 1. Electric engineering—Examinations, questions, etc. I. Keller, K. J. (Kimberly J.) II. Title.
TK169.C65 2007
621.319′24076—dc22

2007019191

Copyright © 2007 by The McGraw-Hill Companies, Inc. All rights reserved. Printed in the United States of America. Except as permitted under the United States Copyright Act of 1976, no part of this publication may be reproduced or distributed in any form or by any means, or stored in a data base or retrieval system, without the prior written permission of the publisher.

McGraw-Hill books are available at special quantity discounts to use as premiums and sales promotions, or for use in corporate training programs. For more information, please write to the Director of Special Sales, Professional Publishing, McGraw-Hill, Two Penn Plaza, New York, NY 10121-2298. Or contact your local bookstore.

    4 5 6 7 8 9 0 QPD/QPD 0 1 3 2 1 0 9

ISBN-13: 978-0-07-148930-0
ISBN-10: 0-07-148930-4

This book was printed on acid-free paper.

Sponsoring Editor Cary Sullivan	**Proofreaders** Lone Wolf Enterprises, Ltd.
Editing Supervisor David E. Fogarty	**Production Supervisor** Pamela A. Pelton
Project Manager Virginia C. Howe	**Composition** Lone Wolf Enterprises, Ltd.
Copy Editor Joseph Staples, Ph.D.	**Art Director, Cover** Jeff Weeks

Information contained in this work has been obtained by The McGraw-Hill Companies, Inc. ("McGraw-Hill") from sources believed to be reliable. However, neither McGraw-Hill nor its authors guarantee the accuracy or completeness of any information published herein, and neither McGraw-Hill nor its authors shall be responsible for any errors, omissions, or damages arising out of use of this information. This work is published with the understanding that McGraw-Hill and its authors are supplying information but are not attempting to render engineering or other professional services. If such services are required, the assistance of an appropriate professional should be sought.

Contents

Acknowledgments and Dedications viii
Preface ix
Understanding the NEC® x

Chapter 1 **ARTICLES 90 THROUGH 110** 1
    Quiz 1—Articles 90 and 100 10
    Quiz 2—Article 110 14
    Quiz 3—True/False 18
    Answers 21

Chapter 2 **WIRING REQUIREMENTS AND PROTECTION** 23
    Quiz 1—Principles-Article 200 28
    Quiz 2—Review-Article 200 31
    Quiz 3—True/False 35
    Quiz 4—Multiple Choice 40
    Answers 49

Chapter 3 **WIRING METHODS AND MATERIALS** 55
    Quiz 1—Article 300 62
    Quiz 2—Comprehensive Test-NEC® Chapters 1 through 3 65
    Quiz 3—True/False 69
    Quiz 4—Multiple Choice 78
    Answers 95

Chapter 4 **EQUIPMENT FOR GENERAL USE** 101
    Quiz 1—Article 400 110
    Answers 117

Chapter 5 SPECIAL OCCUPANCIES AND CLASSIFICATIONS 119
Quiz 1 129
Quiz 2—Classifications 133
Answers 136

Chapter 6 SPECIAL EQUIPMENT 139
Quiz 1 149
Quiz 2 166
Answers 174

Chapter 7 SPECIAL CONDITIONS 177
Quiz 1 183
Quiz 2 186
Answers 197

Chapter 8 COMMUNICATIONS 199
Quiz 1 204
Quiz 2 212
Answers 217

Chapter 9 TABLES, ANNEXES, AND EXAMPLES 219
Quiz 1—Review 236
Quiz 2 238
Quiz 3 245
Answers 253

Chapter 10 MATH CALCULATIONS AND BASIC ELECTRICAL THEORY 257
Timed Math Test 265
Answers 269

Chapter 11 REVIEW AND APPLYING PRINCIPLES 271
Quiz 1—Basic Electrical Principles 281
Answers 292

Chapter 12 MASTER ELECTRICIAN SKILLS 293
Master Level Quiz 298
Answers 303

| Contents | vii |

Chapter 13 TIMED TEST 1 .... **305**
    Timed Test 1 306
    Answers 314

Chapter 14 TIMED TEST 2 .... **317**
    Timed Test 2 319
    Answers 326

Chapter 15 TIMED TEST 3 .... **329**
    Timed Test 3 331
    Answers 347

Chapter 16 CLOSED BOOK EXAM **351**
    Closed Book Exam 352
    Answers 366

Chapter 17 TECHNIQUES FOR STUDYING AND TAKING YOUR TEST . **367**
    Studying for the Exam 368
    Taking Your Exam 369

Acknowledgments and Dedications

I would like to acknowledge my employers. Without their training, support, and the opportunities they have given me, I would not know half of what I know today.

—B. D. Coffin

I would like to dedicate my contributions to this book to my children, Afton and Adam; my shining star and my lighthouse. Their energy, love, and enthusiasm are what make everything I do worthwhile. I would also like to thank Taven for ever reminding me that, in time, anything is possible.

—K. J. Keller

We would both like to thank the National Fire Protection Association for permission to reference the 2005 NEC® standards and for allowing us to reprint materials from the 2005 National Electrical Code® book for the benefit of our readers.

National Electrical Code® and NEC® are registered trademarks of the National Fire Protection Association, Inc., Quincy, MA 02169.

Preface

This book has been written for experienced electricians who plan to take a state required exam in order to obtain a Journeyman or Master electrician's license. We are going on the assumption that you are entering this phase of your career with a basic understanding of the electrical trade and electrical theory. One thing you need to know before you take your licensing exam is that your hands-on experience in the field will not be enough to get you a passing grade.

Many of the questions on licensing exams are based on the National Electrical Code®, so you will need to have a current NEC® book to study and use for reference. Some states will allow you to refer to your code book during the test, while others give "no book" or "closed book" exams. Test centers which allow open book exams will only permit you to bring your NEC® code book, even though a percentage of the test questions will be on electrical theory, controls and other principals used in the safe installation of electrical work. All of the tests are timed and require a testing fee. We know your time and money are valuable, so our goal is to provide you with the study tools, references, and examples you will need to thoroughly prepare yourself to pass your licensing exam.

There are over 1,500 sample test questions presented in this book in a manner which closely resembles the types of questions you may find on the licensing exam. There are True/False and multiple-choice questions, just like on the real test. In order to increase your understanding of the questions, there are answer sheets which include cross-references to the precise article and section of the NEC® from which the question is taken.

The text material at the beginning of each chapter also explains applicable articles and requirements in a manner which makes even the most complicated code standards easier to understand.

There are numerous Test Tips throughout the book which explain how to arrive at the correct answers. Additionally, there are Code Updates in each chapter which clarify changes in the 2005 NEC®, so that you can be certain that the material you are studying is current and accurate. The NEC® covers specialized installation requirements, many of which you may not have performed as an electrician, such as working in hazardous locations, or wiring for hospitals or low-voltage communications systems. Lack of trade experience can be compensated for by studying the NEC® and by taking mock exams. That's why there are practice exams at the end of the book, so that you can get a feel for the type of material you might expect to find when you take your actual licensing exam.

The purpose of this book is to help you develop your ability to understand how licensing exam questions may be written, which standards apply, and how to arrive at the most correct answers in order to receive a passing grade.

B. D. Coffin
K. J. Keller

Understanding the NEC®

When you started working in the electrical industry, all you probably had to do was apply for a helper or associate license and pay a minimal fee to process the application. Now you have worked for thousands of hours and have acquired extensive hands-on experience. You are ready to move up to the next level and become a journeyman electrician, or perhaps its time for you to go into business for yourself and become a master electrician. The first thing you need to do is to contact your state licensing organization and determine the requirements for licensure in your area. Only Illinois, Indiana, Kansas, Kentucky, Missouri, New York, and Pennsylvania do not expect you be licensed at the state level. Most states require that you have completed a specific number of hours of service as an apprentice and some form of study program. Often this means that you cannot even consider getting your license unless you have at least 8000 hours of documented experience and that you have satisfactorily completed a 45-hour current National Electric Code® course. Many locations require completion of state-approved trade-related courses with 225 or more credit hours or graduation from an accredited regional vocational program. Finally, almost every state in the country requires that you take and pass a written test in order to be licensed to perform electrical work without supervision. This test is composed of questions taken directly from and the National Electrical Code®.

The National Fire Protection Association, Inc., known by the acronym NFPA®, develops and publishes the National Electrical Code® (NEC®), which is a compilation of codes and standards for the International Electrical Code. Each section of the NEC® falls under NFPA® 70 and has been developed through a committee process to establish standards for electrical installations to reduce the risk of hazards. The National Electrical Code® is set up by articles that categorize general areas of electrical work, and each article is subdivided into sections that detail standards for specific aspects of that work.

All of the written licensing tests are timed, based on the NEC® codes, and many allow you three hours to complete the questions. The daunting reality is that there are forty-one articles and over 714 pages of code. It would be unrealistic to expect anyone to memorize every article, section, and subsection of the NEC®. Many states allow open-book license testing with very strict guidelines. For example, you can only bring a hardcover version of the NEC® without any tabs or loose papers. This means you can't bring a list of articles or page numbers or any other form of quick reference. Obviously, you don't want to spend any more time than absolutely necessary flipping through your code book searching for answers. You won't be asked to list an article or section on the test—only to know how the code is represented in each question. The exam will be broken down in the following categories:

1. Grounding and bonding
    - System and circuit grounding requirements
    - Grounding methods and locations
    - Proper sizing of grounding conductors
    - Bonding enclosure sizing
    - Equipment and metal-piping-system requirements

2. Services, feeders, branch circuits, and overcurrent protection
    - Electrical loads
    - Proper size and type of service and feeder conductors and ratings
    - Installation of panel boards, switchboards, and overcurrent devices
    - Circuit requirements and applications
    - Electrical outlets, devices, wire connectors, and methods

3. Raceways
    - Types of raceways and uses
    - Proper sizing of conductor fill, supports, and installation methods
    - Type, use, and support methods of boxes, cabinets, etc

4. Conductors
    - Determination of amperage
    - Type of insulation
    - Conductor-usage requirements
    - Methods of installation, protection support, and termination

5. Motors and Controls
    - Installation of generators, motors, and controls
    - Calculations for motor feeders and branch circuits
    - Short-circuit, ground-fault, and overload protection
    - Proper means of disconnection
    - Control-circuit requirements
    - Motor types, applications, and uses

6. General-Use Equipment
    - Lighting and appliances
    - Heating and air-conditioning equipment
    - Generators, transformers, etc.

7. Special Occupancies and Equipment
    - Hazardous locations
    - Health-care facilities, places of assembly, etc.
    - Signs, welders, industrial machinery, swimming pools, etc.

8. General Knowledge of Electrical Trade and Calculations
    - Terminology
    - Practical calculations including load computations and voltage drop

- Conductor derating
- Power factors
- Current ratings of equipment
- Branch circuit calculations

9. Alarm, Communication, and Low-Voltage Circuits
   - Circuits and equipment under 50 volts
   - Signal alarm systems and sound systems

The first four chapters of the NEC® include general electrical standards that apply to all installations. Chapters 5, 6, and 7 cover special applications such as special occupancy, special equipment, and special conditions. Chapter 8 covers communication systems that are not included in the first seven chapters. The last chapter includes reference tables. The Annex section, which follows Chapter 9, provides more tables that illustrate conditions such as ampacity calculations, conduit and tubing fill for conductors and fixture wires, as well as special conditions for conditions from single-family dwellings and industrial applications to park trailers. These examples are particularly helpful, because each case references the applicable code article and section.

Your challenge in preparing to pass the electrical examination will be to gain a clear and basic understanding of the NEC®. If you are taking an open-book exam, you will need to know how to quickly locate areas in the NEC® book that you might need to access for a specific standard so that you can choose the correct answer to the test question. As you go through this workbook, you will have the opportunity to review each code by chapter and test your knowledge of the related codes. Answer sheets for the tests and quizzes provide a reference for each answer so that you can go back over material that you missed.

This is your opportunity to pinpoint your strengths and weaknesses and to improve your knowledge base and test-taking skills. No one wants to take the licensing exam more than once or wait for the next testing date. These are just some of the reasons why this book may prove to be the most valuable tool you own right now. Unlike your wire cutters or your ohmmeter, this tool can help you to get licensed, supervise projects, bid your own work, and earn a higher wage. So what are you waiting for—let's get started.

Chapter 1
ARTICLES 90 THROUGH 110

Electrician's Exam Study Guide

Let's start with the easiest parts of the NEC®—the fist two articles. Article 90 of the NEC® is a basic introduction to intention of the codes. Essentially it describes the purpose of the National Electrical Code, which is to provide uniform and practical means to safeguard people and equipment from electrical hazards. These safety guidelines are not meant to describe the most convenient or efficient installations and don't even guarantee good service or allow for future expansion. They are merely designed to provide a standard for safety that protects against electrical shock and thermal effects, as well as dangerous overcurrents, fault currents, and overvoltage. The NEC® parallels the principles for safety covered in Section 131 of the International Electrotechnical Commission Standard for electrical installations for buildings. The NEC® is divided into nine chapters as illustrated in Figure 1.0 below. This organization builds on the fundamentals in a logical and sequential manner, as you can see in the illustration.

Article 90 provides for special permission from the Authority Having Jurisdiction (AHJ) to grant exception for the installation of conductors and equipment that are outside of a building, or that terminate immediately inside a building wall. The AHJ can also waive certain requirements of the code or may allow alternative methods as long as these alternatives ensure effective safety. Other than these provisions for exception, the code is intended to be a standard that government bodies can enforce. Some of the NEC® rules are mandatory; these can be identified by the terms *shall* or *shall not*. Other rules describe actions that are allowed but are not mandatory; these are characterized by the terms *shall be permitted* or *shall not be required*. Throughout the code book, principles are explained or cross-referenced to related parts of the code in the form of Fine Print Notes (FPN); these notes are merely informational and are not enforceable.

The rest of Article 90 covers wiring planning in general terms. It explains that metric units of measurement (SI) are listed first with inch-pound units following and that trade practices are used in trade sizing. There were a number of wording changes to Articles 90 and 100 in the 2005 NEC®. These alterations are incorporated into the various Code Update courses offered around the country and are included on the li-

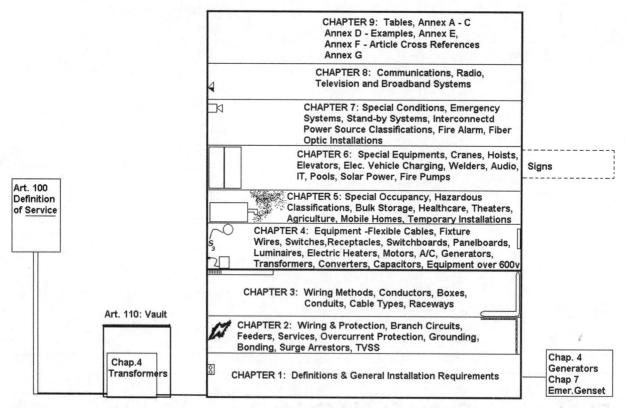

FIGURE 1.1 NEC® codes ascend in a logical manner, building up from the most fundamental aspects to more complex and specialized installations.

censing exam. Once you have your license, it may be necessary for you to take periodic update programs in order to renew your license and to keep current on the most recent code requirements. Several of the changes to Articles 90 and 100 are listed for you in this chapter.

> **! Code update ARTICLE 90**
>
> Signaling and communications conductors, equipment and raceways, as well as fiber optic cables and raceways are also included in the scope of the 2005 NEC® if they are installed in the locations and occupancies listed in 90.2(A)(1)&(2).

> **! Code update SECTION 90.2 (B)**
>
> According to 90.2(b) Any installation under the exclusive control of the utility is not covered by the NEC. However, in some large facilities like industrial plants, institutions or corporate office complexes, the customer owns and maintains the generators, transmission and distribution lines. In such cases, the Authority Having Jurisdiction may determine the National Electrical Safety Code, and not the National Electrical Code, is the standard which applies to this type of wiring.

> **! Code update SECTION 90.9(C)(4)**
>
> 90.9(C)(4)–There are two ways to convert from inch-pound to metric measurements. (1) Hard conversion and (2) Soft conversion. A soft conversion is more accurate than a hard conversion. For example, the soft conversion for 3 ft. is 914.4 millimeters. The hard conversion is 900 millimeters.
>
> When the conversion from a metric number to an inch-pound measurement involves a safety issue, like in the approach distances to exposed live parts, a soft or exact conversion must now be used.

Chapter One of the NEC® begins with Article 100, which is an alphabetical listing of definitions that are key to the proper application of the code. Some of these definitions are simple enough. For example, "Dwelling, One-Family" is described as a "building that consists solely of one dwelling unit." Many of the descriptions are written in more complex terms. An outlet, for example, is defined as "a point on the wiring system at which current is taken to supply utilization equipment." A "power outlet" is "an enclosed assembly that may include receptacles, circuit breakers, fuseholders, fused switches, buses, and watt-hour meter mounting means; intended to supply and control power to mobile homes, recreational vehicles, park trailers, or boats or to serve as a means for distributing power required to operate mobile or temporarily installed equipment." What did that just say? You've worked in the electrical industry for quite a while. Is that how you would have described a power outlet? If you had been given that definition and the choice of Outlet, Power Outlet, and Receptacle, would you know which answer was correct?

There are several keys to unlock such difficult definitions. First, look for key words within the definition. In the "Power Outlet" example above, you can see that the item described can include receptacles. You immediately know the answer is not "Receptacle" because the definition will not use the name of the device to describe itself. This step reduces your possible answers to two. At this point, you could either guess at the correct choice, knowing you have a fifty-fifty chance of getting the right answer, or you could look for more clues in the description. When you look at the definition again, you'll see that this item is intended to supply and control "power" and to distribute "power." This makes it clear that power is an essential part of the answer; therefore, "Power Outlet" makes the most sense. Of course, if it's an open-book exam, you could grab your NEC® book and flip to the definition section in Article 100 to look up the answer, but whenever possible, try not to use up valuable time going through your code book. Use your powers of reason and deduction first, and save your code book for the really complex subjects so that you do not run out of test time.

> ▶ **Test** tip
> One of the sections on the exam covers General Knowledge of Electrical Trade and Calculations—Terminology. Article 100 is an alphabetical listing of electrical terms and definitions.

Another essential key to passing your exam is to remember that terms are not interchangeable and they mean different things. We learned in elementary school that the word "read" means two different things depending on how it is used in the sentence. You can *read* this book, and once you've finished, you have *read* the book. Article 100 has similar occurrences. For instance, what is the difference between a grounded conductor and a grounding conductor? These two terms are often used synonymously, but the NEC® defines them in completely different ways. A grounded conductor is a circuit conductor that is intentionally grounded. A grounding conductor is used to connect equipment or a grounded circuit to a grounding electrode. Knowing that there is difference between the two terms, and what that variation is,

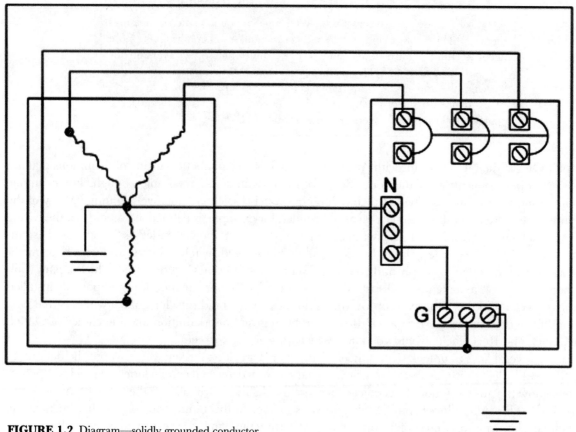

FIGURE 1.2 Diagram—solidly grounded conductor.

will be the difference between getting an answer correct and having it count towards your final score and answering a question wrong and having it basically mean nothing.

As you study, try to keep a couple of things in mind. First of all, the exam is multiple choice. Some questions may be True/False, while others will list several answer options. The example of the Power Outlet above might make you think that the exam is written to intentionally confuse you, but it isn't. The NEC uses expert-level, specialized vernacular, and so the exam questions may seem complicated and hard to understand. Bear in mind that the NEC® is not just used by electricians. Electrical inspectors, engineers, designers, and architects also use the standards established in the NEC. You may need to read the question carefully before you can understand the meaning of the question and determine which answer option is the most correct. If you need to, circle key words or values in the example so that you can break the question down into clearer components.

> **! Code update ARTICLE 100 - DEFINITIONS**
>
> ***Coordination:*** The definition of coordination has been moved from Article 240, Overcurrent Protection, to Article 100 because the new definition is used in more than a single Code Article.

The next thing to consider is that only the correct answers count towards your score, and incorrect answers don't count against you. This means it is in your best interest to answer every question. Obviously you want to answer as many questions correctly as possible because in most cases you will have to get a minimum score of 70% to receive a passing grade. Each chapter of this book includes a Quick Quiz or sample test on the sections covered in the chapter so that you can assess not only your knowledge of the material, but also your ability to achieve a high percentage of correct answers. You will find *Test Tips* throughout the book to help you decipher questions which may seem complicated or just plain tricky, and *Code References* have been included in the answer sections so that you can go directly to the NEC® for a clearer understanding of test answers. Your overall goal is to increase the number of correct answers you score by combining your knowledge, skills, and ability to arrive at the most correct answer during a test.

> **! Code update ARTICLE 100 - DEFINITIONS**
>
> ***Grounded, solidly:*** Originally found in Article 250.95, this definition has been changed slightly and moved to Article 100 because when a term is used in two or more Articles it belongs in Article 100. The new definition says Solidly Grounded means "Connected to ground without inserting any resistor or impedance device."

> **! Code update ARTICLE 100 - DEFINITIONS**
>
> ***Device:*** The new definition of a device says a device can "carry or control but not utilize electric energy." The old definition said a device could "carry but not utilize electric energy."

! Code update ARTICLE 100 - DEFINITIONS

Grounding Electrode: The different types of grounding electrodes are listed in 250.52, but in the 2005 NEC® the definition has been included in Article 100. A grounding electrode is defined as "A device that establishes an electrical connection to the earth."

! Code update ARTICLE 100 - DEFINITIONS

Guest Room; Guest Suite: Guest rooms and guest suites are "compartments" within hotels and motels. "Guest Suites" was added because many hotels have accommodations with more than a single room. A Guest Suite is defined in the NEC® as: "An accommodation with two or more contiguous rooms comprising a compartment, with or without doors between such rooms, that provides living, sleeping, sanitary, and storage facilities." The definition does not include "permanent provisions for cooking," so guest rooms and guest suites are not dwelling units. If, for example a kitchenette is included in a guest room or suite, then it becomes a dwelling unit and all the requirements for circuits and outlets in dwelling units apply.

! Code update ARTICLE 100 - DEFINITIONS

Handhole Enclosures: The new definition of handhole enclosure explains underground enclosures and emphasizes they must be accessible from the surface so personnel can reach into the enclosure. A handhole enclosure which is suitable for use by power conductors must be listed for that use. Handhole enclosures listed for use with communications cables cannot be used for lighting and power wires.

! Code update ARTICLE 100 - DEFINITIONS

Outline Lighting: The old definition was an arrangement of incandescent lamps or electric-discharge lighting to outline or call attention to certain features such as the shape of a building or the decoration of a window. In the 2005 NEC, the definition was altered to read, "An arrangement of incandescent lamps, electric discharge lighting, or other electrically powered light sources to outline or call attention to certain features such as the shape of a building or the decoration of a window." The new definition allows different light sources, like low-voltage light emitting diodes (LEDs), to be used as outline lighting.

> **! Code update ARTICLE 100 - DEFINITIONS**
>
> ***Qualified Person:*** A new Fine Print Note (FPN) was added in the 2005 NEC® under Qualified Person which says: "Refer to NFPA® 70E-2004 for electrical safety training requirements."

> **! Code update ARTICLE 100 - DEFINITIONS**
>
> ***Separately Derived System:*** The previous definition of a separately derived system included a list of possible energy sources such as batteries, photovoltaic systems, generators, transformers and converters. The new definition eliminated the list and defines a separately derived system as "A premises wiring system whose power is derived from a source of electric energy or equipment other than a service."

> **! Code update ARTICLE 100 - DEFINITIONS**
>
> ***System Bonding Jumper:*** The main bonding jumper is the "connection between the grounded circuit conductor and the equipment grounding conductor at the service." The system bonding jumper is the "connection between the grounded circuit conductor and the equipment grounding conductor at a separately derived system." The new definition simply clarifies that a bonding connection must be made between the grounded (neutral) conductor and the equipment grounding conductor at a transformer, just like at a service.

There are a number of key points covered in Article 110. Keep in mind that none of the standards in this article apply to communications circuits; however, some of the principles are referenced in Chapters 6 and 7 of the NEC® for wiring systems. Article 110 lays the ground work for overall installation requirements. Throughout this book, article sections and sub-sections will be referenced and identified in bold print brackets such as **[110.3(B)]**

The Authority Having Jurisdiction (AHJ) has to approve all electrical equipment used, including low-voltage and limited-energy cables and equipment. Article 100 defines what this approval entails.

Article 110 covers general requirements for electrical installations. Sections **[110.5]** and **[110.6]** explain conductors and conductor sizes respectively. The key points are that conductors discussed in the Code are copper, except for the aluminum and copper-clad aluminum conductors covered in **[310.5]**, and that conductor sizes are measured in American Wire Gage (AWG).

Section **[110.12]** explains requirements for the mechanical execution for electrical installations. For example, **[110.12(A)]** states that if metal plugs or plates are used with non-metallic enclosures, then they must be recessed ?" from the outer surface of the enclosures. Unused raceway or cable openings in boxes and conduits must also be closed so that the protection provided is at least equal to the protection provided by the wall of the box or conduit. The bottom line? Knock-out plugs must be used for any unused openings in a box.

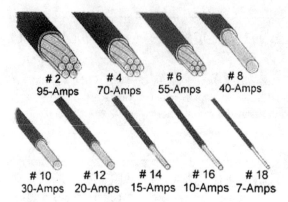
FIGURE 1.3 AWG conductor sizes.

Some of the section regards common sense elements of installation, such as ensuring that materials are not damaged or contaminated by materials that are abrasive or corrosive. Equipment provisions for terminations of equipment with circuits with 100 amperes or less or 14 AWG to 1 AWG conductors are detailed in section **[110.14 (C)(1)(a)]** and **[110.14(C)(1)(b)]**. The bottom line of these sections is that circuits and conductors within these size ranges can only be used for conductors rated 60 degrees C (140 degrees F) and motors with design letters B, C, or D with conductors rated 75 degrees C (167 degrees F). Any exceptions are also outlined in these sections.

Section **[110.16]** describes in general terms the need to label non-residential equipment such as switchboards, panel boards, and control panels with any potential electric arc flash hazards. A note in this section refers you to NFPA® 70E for details on arc flash protection and calculations. One thing you should be aware of is that, while NFPA® 70E has not yet been adopted as part of the electrical Code, inspection officers around the country are being trained to enforce the principals of NFPA® 70E. We have provided more details on this provision in the Appendix section of this book.

Section **[110.26(A)(1)]** covers the depth of work space involving live parts and includes a table for illustration. Three separate conditions are listed and the established safe distance for each is provided. Three exceptions to the work space regulation, such as dead-front assemblies, special permissions for low voltage, and existing conditions are listed in **[110.26(A)(1)(a)]** through **[110.26(A)(1)(c)]**. Height, width, and headroom considerations, clear spaces, as well as indoor and outdoor installations are covered in the remainder of section **[110.26]**.

> **! Code update SECTION 110.26**
>
> *110.26 Spaces About Electrical Equipment:* The previous code wording stated that "an enclosure was accessible if it was controlled by "lock and key." The 2005 NEC® changes the wording so enclosures are now considered accessible if they are controlled by "lock(s)." This allows any type of lock. Locks with keys, combination locks, keypad locks or electric locks are permissible as long as qualified individuals have the means to unlock the enclosure.

Section **[110.26]** covers conductors and equipment using circuits over 600 volts nominal, and the minimum distance from a fenced enclosure to any live parts is listed in Table **[110.31]**. Issues concerning fire resistance, indoor and outdoor installations, and requirements for enclosures that are accessible to unqualified people are covered in the balance of this section. Section **[110.34]** outlines work space specifications and guarding of equipment. The minimum depth of clear working space and the various conditional requirements are illustrated in Table **[110.34(A)]**, and the mandatory elevation of unguarded live parts above a working space is listed in Table **[110.34(E)]**.

> **! Code update SECTION 110 - TABLE 110.26 (A)(1)**
>
> *Table 110.26(A)(1) Working Space:* The wording in condition 2 has been changed to make it clearer that condition 2 applies if a grounded surface is on the other side of the work space from the exposed, energized parts. In the 2002 NEC® it was not clear whether the grounded surface referred to the surface where the equipment was mounted, or the opposite surface on the other side of the work space.

Part IV of Article 110 covers installations of equipment, over 600 volts nominal, in tunnels. High-voltage conductor tunnel installations require metal conduit or metal raceways and Type MC cable as explained in **[110.53]**. Keep in mind that Article 110 deals mostly with general installation requirements. Throughout the Article are cross-references to other sections of the Code which approach equipment installations in more specific terms. If you read part V of Article 110, you will be introduced to manholes and other enclosures of all voltages. For example, section **[110.53]** states that these enclosures must be big enough to permit "ready" withdrawal and installation of conductors without causing the equipment any damage. Remember that while section 100 gives some specific details for installations, such as enclosure size requirements, most of this part of the Code concerns overall safety guidelines and industry common sense.

Now it's time to speed test your knowledge of the information covered in Articles 90 through 110. Remember, when you take the quizzes and tests in this book, always answer all of the questions and choose the most correct, or "best," answer to the questions. At this point you are just learning how to read and interpret the questions; you are not trying to answer the questions within a set period of time.

QUIZ 1—ARTICLES 90 AND 100

1. Installations of electrical equipment in ships, railway vehicles, and aircraft are covered by the NEC.
   - **a.** True
   - **b.** False

2. Compliance with the NEC® shall result in installations which are efficient, adequate for good service, and free from hazard.
   - **a.** True
   - **b.** False

3. The Code may require new products, materials or constructions that may not yet be available at the time the Code is adopted.
   - **a.** True
   - **b.** False

4. The Code is written in an explicit manner and may be used as an instructional manual for untrained persons.
   - **a.** True
   - **b.** False

5. Installations of electrical equipment in recreational vehicles, mobile homes and floating buildings are covered by the NEC.
   - **a.** True
   - **b.** False

6. Limiting the number of circuits in a single enclosure does not minimize the effects of a ground fault in one circuit.
   - **a.** True
   - **b.** False

7. The use of either numbers shown in the SI system or in the inch-pound system shall constitute compliance with the Code.
   - **a.** True
   - **b.** False

8. A reliable conductor used to ensure the required conductivity between metal parts which are to be electrically connected is called which of the following:
   - **a.** Connector
   - **b.** Bonding jumper
   - **c.** Insulated conductor
   - **d.** Branch connector

9. The current measured in amperes which a conductor can carry continuously under standard conditions without exceeding the conductor's maximum temperature rating is called:

   a. Maximum current
   b. Conductor current
   c. Ampacity
   d. Convection amps

10. Limiting the total quantity of circuits in any given enclosure does which of the following:

    a. Minimizes the effects of a short circuit in one circuit
    b. Eliminates over-expansion in the future
    c. Provides easy access to qualified workers
    d. Provides a standard for the Authority Having Jurisdiction

11. Items in the Code that identify actions that are specifically required or prohibited are considered which of the following:

    a. Permissive rules
    b. Mandatory rules
    c. Guidelines for examination of installation
    d. Permitted use

12. Two or more ungrounded conductors with equal voltage running between them and a grounded conductor is considered which of the following:

    a. Control circuit
    b. Feeder
    c. Branch circuit
    d. Loop feed

13. A device that de-energizes a circuit or a portion of a circuit within an established time frame when a current to ground exceeds the values determined for a Class A device is which of the following:

    a. Circuit breaker
    b. Fuse
    c. Ground-fault interrupter
    d. Voltage regulator switch

14. A device that protects a motor against overheating and is an integral part of that motor is considered which of the following:

    a. Integral fuse
    b. Ground-fault protector
    c. Thermal protector
    d. Shunt trip switch

15. An assembly of two or more single-pole fuses is which of the following:

    a. Multiple fuse
    b. Multi-tap connector
    c. Switching device
    d. Panelboard

16. A continuous load is one in which one of the following types of current is expected to run non-stop for three hours or more:
    a. Continuous
    b. Normal
    c. Maximum
    d. Limited

17. A branch circuit where two or more ungrounded conductors, with a potential difference between them, and a grounded conductor with equal potential difference between it and each ungrounded conductor is referred to as which of the following:
    a. Continuous loop feed system
    b. Multi-wire
    c. General purpose circuit
    d. Regulated branch circuit

18. The purpose of the National Electrical Code is to provide which of the following:
    a. A standard by which professionals may design installations
    b. Efficient electrical installations
    c. Cost-effective installations
    d. Safe electrical installations

19. When a single piece of equipment is within sight of another piece of equipment, the pieces must be within how many feet of each other?
    a. 10
    b. 25
    c. 40.5
    d. 50

20. Circuit voltage is best defined as which of the following:
    a. Average potential between two conductors
    b. Maximum potential difference between two conductors
    c. Effective difference of potential between two conductors
    d. Total amperes produced between two conductors

21. The connection device installed at an outlet to allow for two or more contact devices at the same yoke is which of the following:
    a. Duplex outlet
    b. Multiple receptacle
    c. Slice connector
    d. None of the above

22. A large single panel assembly of panels containing mounted switches, overcurrent and protection devices and buses is considered which of the following:
    a. Panelboard
    b. Switchboard
    c. Automatic transfer switch
    d. Service panel

Articles 90 Through 110   13

23. A switching device utilized to isolate a circuit or equipment from an established power source is determined to be which of the following:

    a. Interrupter switch   **b.** Circuit breaker

    c. Cutout   **d.** Disconnection Switch

24. A fuse may or may not be the complete device necessary to connect it to an electrical circuit.

    a. True   **b.** False

25. A compartment to which one or more air ducts are connected to form part of an air distribution system is called which of the following:

    a. Plenum chamber   **b.** Ventilation duct

    c. Air-flow box   **d.** Circulation cavity

26. Which of the following installations are covered by the 2005 National Electrical Code:

    a. Power wiring for a large industrial machine

    b. Computer wiring in an office building

    c. Telephone cable in a flexible duct

    d. All the above

27. The National Electrical Code does not apply to:

    a. City-owned overhead distribution lines

    b. Utility-owned transmission lines

    c. Transformers serviced by utility technicians

    d. All the above

28. Which of the following statements is considered mandatory in the NEC:

    a. See UL Standard 1699 on Arc-Fault Circuit Interrupters

    b. [NFPA® 70e]

    c. FPN: For conductors exposed to deteriorating agents, see 310.9

    d. Circuits and equipment shall comply with 820.3(A) through 820.3(G)

29. A soft conversion from inch-pounds to metric measurements is which of the following:

    a. More accurate   **b.** Less accurate

    c. Easier to calculate   **d.** Harder to determine

30. In a wiring system that is engineered for selective coordination, which of the following conditions must exist:

   a. The upstream overcurrent device shall be less than twice the rating of the downstream overcurrent device.

   b. The upstream overcurrent device contains at least one fast-acting fuse and the downstream overcurrent device is a circuit breaker.

   c. The downstream device will trip before the upstream device trips.

   d. The downstream device will allow an overcurrent condition above its rating to be cleared by the upstream overcurrent device.

QUIZ 2—ARTICLE 110

1. Chose the answer below that is not an approved means of mounting electrical equipment to a masonry wall:

   a. Using screws that are driven into wooden plugs in the wall

   b. Using bolts that are supported by metal plates on the back side

   c. Using molly bolts through holes drilled completely through the wall

   d. Using lag bolts screwed into lead masonry anchors

2. The clear work space on one side of equipment with exposed live parts, which operates at 600 volts nominal or less to ground, and which will require examination, adjustment or maintenance shall be at least which of the following:

   a. 12 inches wide

   b. 24 inches wide

   c. 30 inches wide

   d. 2 times the width of the panel

3. The floor of a vault for electrical equipment with either a vacant space or additional floors below it shall have a minimum fire resistance of which of the following:

   a. 1 hour
   b. 3 hours
   c. 6 hours
   d. 12 hours

4. Circuits with 100 amperes or less, or with 14 AWG to 1 AWG conductors, can only be used for conductors rated at which of the following:

   a. 55 degrees C (131 degrees F)

   b. 60 degrees C (140 degrees F)

   c. 75 degrees C (167 degrees F)

   d. 80 degrees C (176 degrees F)

5. Voltage to ground of 0-150, with exposed live parts on one side of the work space and no live or grounded parts on the other side, must have a minimum clear distance of which of the following:

   a. One foot
   b. Two feet
   c. Three feet
   d. 42 inches

6. The work space for equipment must allow 90-degree opening of any equipment doors or hinged panels.

   a. True
   b. False

7. The minimum clear headroom for work space around service equipment, switchboards, panelboards, or motor control centers must be which of the following:

   a. 3 feet
   b. 4 1/2 feet
   c. 6 1/2 feet
   d. 10 feet

8. The minimum clear working-space depth for equipment over 75 kV nominal volts to ground with exposed live parts on both sides is which of the following:

   a. 3 feet
   b. 6 feet
   c. 10 feet
   d. 12 feet

9. High-voltage conductor tunnel installations require metal conduit or metal raceways and which of the following:

   a. Type MC cable

   b. Aluminum conductors

   c. Copper-clad aluminum conductors

   d. EMT cable

10. Unused raceway or cable openings in boxes and conduits must be closed so that the protection provided is which of the following:

    a. At least equal to the protection provided by the wall of the box or conduit

    b. Equal to the depth of the wall framing

    c. Greater than the protection provided by the box or conduit alone

    d. Adequate to act as a fire stop

11. The elevation for unguarded live parts in a working space for equipment operating between 601 to 7500 nominal volts between phases is which of the following:
    a. 3 feet
    b. 6 feet
    c. 9 feet
    d. 12 feet

12. Voltage to ground of 150-600, with exposed live parts on one side of the work space and grounded parts, such as concrete, brick or tile walls, on the other side must have a minimum clear distance of at least which of the following:
    a. 12 inches
    b. 24 inches
    c. 36 inches
    d. 42 inches

13. The fire rating for walls, floors and doors containing equipment over 600 volts nominal shall be a minimum of which of the following:
    a. 1 hour
    b. 2 hours
    c. 3 hours
    d. 6 hours

14. Often, equipment and terminations are labeled with which of the following:
    a. The initials of the installer
    b. Service tag
    c. Tightening torque
    d. Wiring designations

15. Ventilation system electrical controls shall be installed in manner in which the airflow can be managed in which of the following ways:
    a. Vented to the outside
    b. Reversed
    c. Limited upon demand
    d. Prevented

16. If the cover for an underground pull box used for circuits over 600 volts is not locked, bolted, or screwed into place, then it must weigh equal to which of the following:
    a. 25 pounds
    b. 50 pounds
    c. 75 pounds
    d. 100 pounds

17. Entrances to guarded locations containing exposed live parts shall be marked with warnings that are which of the following:
    a. Conspicuous
    b. Universal
    c. At least 6 inches square
    d. Red in color

18. In order to be electrically secure prior to soldering, splices must be which of the following:
    a. Sanded
    b. Joined mechanically
    c. Free of rough edges
    d. Coated with flux

19. Unless otherwise specified, live parts for electrical equipment operating at 50 volts or more shall be guarded.

    a. True   **b.** False

20. Voltage to ground of 150-600 with exposed live parts on both sides of the working space must have a clear distance not less than which of the following:

    a. 2 feet   **b.** 2 1/2 feet

    c. 3 feet   **d.** 4 feet

21. When mounting electrical equipment, wooden plugs driven into plaster may be used.

    a. True   **b.** False

22. Concrete and brick walls are considered which of the following:

    a. Dry locations   **b.** Insulators

    c. Wet locations   **d.** Grounded

23. If a conductor material is not specified in a particular Article or Section, the material shall be assumed to be which of the following:

    a. Copper   **b.** Aluminum

    c. Copper-clad aluminum   **d.** No assumptions shall be made

24. A high-let conductor in a three-phase, 4-wire delta secondary shall be which of the following colors:

    a. White   **b.** Orange

    c. Green   **d.** Black

25. Conductor sizes are listed in which of the following:

    a. Circular mils   **b.** Diameter or thickness

    c. AWG or millimeters   **d.** AWG or circular mils

QUIZ 3—TRUE / FALSE

1. Multiple wiring methods may be used within a single wiring system.
   - **a.** True
   - **b.** False

2. Unused raceway openings in auxiliary gutters must be effectively closed.
   - **a.** True
   - **b.** False

3. Electrical tape may be used to cover splices and the free ends of conductors.
   - **a.** True
   - **b.** False

4. Panelboards and distribution boards may occupy the same space.
   - **a.** True
   - **b.** False

5. An unused current transformer that is associated with potentially energized circuits must have an independent grounding connection.
   - **a.** True
   - **b.** False

6. The minimum distance permitted from a fence to enclosed equipment operating at over 600 volts nominal is 5 feet.
   - **a.** True
   - **b.** False

7. Entrances to rooms containing equipment with exposed live parts shall be marked with warning signs prohibiting unauthorized personnel from entering.
   - **a.** True
   - **b.** False

8. Work spaces around enclosed electric equipment must have access that is at least 2 feet wide by 6 feet high.
   - **a.** True
   - **b.** False

9. A screen is an acceptable means of separating high-voltage equipment from equipment that operates under 600 volts.
   - **a.** True
   - **b.** False

10. In a manhole with equipment that has cables located on both sides, a minimum clear work space of five feet is required.
    - **a.** True
    - **b.** False

11. A branch circuit supplies two or more receptacles for lighting or appliances.
    - **a.** True
    - **b.** False

12. The power factor is a ratio of the maximum demand of part of a system to the total connected load of that part of the system.
    - **a.** True
    - **b.** False

13. Wires that run behind a finished wall are considered to be concealed.
    - **a.** True
    - **b.** False

14. A continuous load is expected to run at maximum current non-stop for 10 hours.
    - **a.** True
    - **b.** False

15. A guest room contains living, sleeping, sanitary, and storage facilities:
    - **a.** True
    - **b.** False

16. An elevator shaft is considered a cableway.
    - **a.** True
    - **b.** False

17. Energized conductive elements or components are considered live parts.
    - **a.** True
    - **b.** False

18. A string of indoor lights that is suspended between two points is called festoon lighting.
    - **a.** True
    - **b.** False

19. An electrical installation constructed to prevent beating rain from interfering with the safe and successful operation of an apparatus is considered to be rainproof.
    - **a.** True
    - **b.** False

20. An electric circuit that controls another circuit through a relay is a remote-control circuit.
    - **a.** True
    - **b.** False

21. Conductors from the service point to the service disconnecting point that are made up in the form of a cable are called service conductors.
    - **a.** True
    - **b.** False

22. The written consent of the Authority Having Jurisdiction is considered approved consent.

   a. True　　　　　　　　　　**b.** False

23. An electrical installation which is constructed to protect an apparatus or connection from exposure to the weather is considered to be watertight.

   a. True　　　　　　　　　　**b.** False

24. Mandatory rules within the NEC® are easily identified by the use of the words "shall" and "shall not."

   a. True　　　　　　　　　　**b.** False

25. A device that governs the electric power delivered to an apparatus to which it is connected in a pre-determined manner is called a controller.

   a. True　　　　　　　　　　**b.** False

ANSWERS

QUIZ 1—ARTICLES 90 AND 100

1. **B** Article 90.2 (B)(1)
2. **B** Article 90.1 (B)
3. **A** Article 90.4
4. **B** Article 90.1 (C)
5. **A** Article 90.2 (A)(1)
6. **B** Article 90.8 (B)
7. **A** Article 90.9 (D)
8. **B** Article 100
9. **C** Article 100
10. **A** Article 90.8 (B)
11. **B** Article 90.5 (A)
12. **C** Article 100
13. **C** Article 100
14. **C** Article 100
15. **A** Article 100
16. **C** Article 100
17. **B** Article 100
18. **D** Article 90.1 (A)
19. **D** Article 100
20. **C** Article 100
21. **B** Article 100
22. **B** Article 100
23. **D** Article 100
24. **A** Article 100 II Fuse FPN
25. **A** Article 100
26. **D** Article 90.2 (A)(1)
27. **D** Article 90.2 (B)
28. **D** Article 90.5 (C)
29. **A** (NOTE: Soft conversions are direct mathematical conversions. Hard conversions are rounded off.)
30. **C** Article 100 Coordination (Selective)

QUIZ 2—ARTICLE 110

1. **A** Article 110.13(A)
2. **C** Article 110.26(A)(2)
3. **B** Article 110.31(A)
4. **B** Article 110.14(C)(1)(a) and 110.14(C)(1)(b)
5. **C** Article 100.26(A)(2)
6. **A** Article 100.26(A)(2)
7. **C** Article 110.34(A)
8. **D** Article 110.34(A)
9. **A** Article 110.53
10. **A** Article 110.12(A)
11. **C** Article 110.34(E) Table
12. **D** Article 110.26(A)(1)
13. **C** Article 110.30(A)
14. **C** Article 110.14 FPN
15. **B** Article 110.57
16. **D** Article 110.31(D)
17. **A** Article 110.27(C)
18. **B** Article 110.14(B)
19. **A** Article 110.27(A)
20. **D** Article 110.26(A)(1)
21. **B** Article 110.13(A)
22. **D** Article 110, Table 110.26(A)(1), condition 2
23. **A** Article 110.5
24. **B** Article 110.15
25. **D** Article 110.6

QUIZ 3—TRUE / FALSE

1. B	8. A	14. B	20. A
2. A	9. A	15. A	21. B
3. B	10. B	16. B	22. B
4. B	11. A	17. A	23. B
5. B	12. B	18. B	24. A
6. B	13. A	19. A	25. A
7. A			

Chapter 2
WIRING REQUIREMENTS AND PROTECTION

Chapter two of the NEC® outlines wiring requirements and usage, as well as protection specifications. Branch circuits, feeders, service wiring, overcurrent protection, grounding and bonding, surge protection and transient voltage surge suppression (TVSS) are detailed in this chapter. These are all fundamental elements of electrical work, so you should be very clear on the requirements covered in Chapter two. The progression of the chapter essentially follows the progression of an installation, and looks like this:

- Identification and use of grounded conductors—Article 200
- Branch circuits—Article 210
- Feeders—Article 215
- Branch circuit, feeder and service calculations—Article 220
- Service installations—Article 230
- Overcurrent protection, so the whole project doesn't blow up—Article 240
- Grounding, so people don't blow up—Article 250
- Surge arrestors—Article 280
- TVSS installations—Article 285

Chapter two of the NEC® is basically the "what to use, how, and when" chapter for power installations under standard, normal conditions. It is all about colors, sizes, current loads, identification and protection. For example, color requirements and other approved means of identifying grounded conductors are listed in Section **[200.6(A)(E)]**. Requirements for branch circuits with combination loads are explained in Article 210, and Table **[210.2]** references articles and sections throughout the NEC® that cover specific-purpose branch circuits. Sections **[210.19(A)]**, **[215.2]** and **[230.42(A)]** specify that the conductor must be sized no less than 100% of the noncontinuous load, plus 125% of the continuous load.

Table 210.24 Summary of Branch-Circuit Requirements

Circuit Rating	15 A	20 A	30 A	40 A	50 A
Conductors (min. size):					
Circuit wires[1]	14	12	10	8	6
Taps	14	14	14	12	12
Fixture wires and cords — see 240.5					
Overcurrent Protection	15 A	20 A	30 A	40 A	50 A
Outlet devices:					
Lampholders permitted	Any type	Any type	Heavy duty	Heavy duty	Heavy duty
Receptacle rating[2]	15 max. A	15 or 20 A	30 A	40 or 50 A	50 A
Maximum Load	15 A	20 A	30 A	40 A	50 A
Permissible load	See 210.23(A)	See 210.23(A)	See 210.23(B)	See 210.23(C)	See 210.23(C)

[1]These gauges are for copper conductors.
[2]For receptacle rating of cord-connected electric-discharge luminaires (lighting fixtures), see 410.30(C).

Reprinted with permission from NFPA® 70, *National Electrical Code*®, Copyright ©2004, National Fire Protection Association, Quincy, MA 02169. This reprinted material is not the complete and official position of the National Fire Protection Association on the referenced subject which is represented only by the standard in its entirety.

Wiring Requirements and Protection

> **! Code update SECTION 215.12**
>
> Ungrounded feeder conductors must be separately identified when there is more than one nominal voltage system in the building. For example, in a commercial building with a 480/277 service and loads operating at 208/120 volts, all of the ungrounded conductors for both systems would need to be separately identified. Identification can include separate color coding, marking tape, tagging, or other approved means. If color coding is used, no specific colors are mandated.

NEC® Section **[240.3]** requires that the branch circuit, feeder, and service conductors be protected against overcurrent based on their ampacities as specified later in Chapter three, Table **[310.16]**. However, Section **[240.3]** contains twelve rules that modify the general requirements and permit the conductors to be unprotected in accordance with their ampacities. These rules include the following:

1. Power loss hazard
2. Devices rated 800 amperes or less
3. Tap conductors
4. Motor-operated appliance circuit conductors
5. Motor and motor-control circuit conductors
6. Phase converter supply conductors
7. Air-conditioning and refrigeration equipment circuit conductors
8. Transformer secondary conductors
9. Capacitor circuit conductors
10. Electric welder circuit conductors
11. Remote-control, signaling, and power-limited circuit conductors
12. Fire alarm system circuit conductors

All conductors must be protected against overcurrent in accordance with Section **[240.3]**. Connection must be made between equipment grounding conductors and any metal box by using a listed grounding device or grounding screw that doesn't serve any other purpose as stated in Section **[240.3]**. This requires that the branch circuit, feeder, and service conductors be protected against overcurrent in accordance with their ampacities as specified in Table **[310.16]**. However, Section **[240.3(B)]** permits "the next size up device" if (1) the conductors are not part of a multioutlet branch circuit supplying receptacles, and (2) the ampacity of the conductors does not correspond with the standard ampere rating of an overcurrent protection fuse or a circuit breaker as listed in Section **[240.6(A)]**, and (3) the next higher standard rating selected does not exceed 800 amperes. Section **[240.6(A)]** contains the list of standard size overcurrent protection devices.

> **! Code update SECTION 240.5**
>
> The 2002 NEC® listed the minimum size cord for listed appliances and portable cords on standard branch circuit sizes. The 2005 edition has deleted that section and replaced it with the requirement that these cords must be protected when applied within the listing requirements of the appliance or portable lamp.

According to Section **[240.10]**, supplementary overcurrent protective devices used in light fixtures, or luminaries, do not have to be readily accessible. Anytime equipment grounding is required, it must be in accordance with Article 250. Also, grounding conductors must not be connected to enclosures with sheet metal screws per Section **[250.8]**. Lighting fixtures and equipment are considered grounded when they are mechanically connected to an equipment grounding conductor as explained in Section **[250.118]**.

Flexible metal conduit (FMC) is approved as a grounding method in Section **[250.118 (5)]** if the following conditions exist: (1) the conduit is terminated in fittings listed for grounding; (2) the circuit conductors contained in the conduit are protected from overcurrent devices rated at 20 amperes or less; (3) the total length in any ground return path is 6 feet or less; and (4) the conduit is not installed for flexibility. Additionally, if an equipment bonding jumper (the connection between two or more sections of an equipment grounding conductor—Article 100) is required around FMC, then it must be installed as outlined in Section **[250.102]**. Section **[250.118(6)]** goes on to outline the use of Liquid tight Flexible Metal Conduit (LFMC) as a grounding means based on the size of the LFMC and the amperes involved. A raceway may be used as an equipment grounding conductor as stated in Section **[250.134(A)]** and in accordance with Sections **[250.4(A)(5)]** or **[250.4(B)(4)]**. EMT can serve as an equipment grounding connector per Section **[250.118(4)]** and they must be sized according to Section **[250.122]**. The minimum size for equipment grounding conductors, grounding raceways, and equipment listed by AWG is found in Table **[250.122]**.

> **! Code update 250.184(B)**
>
> Single-point grounded systems have been added to the 2005 NEC. Single-point grounding is allowed at the source of a separately derived system with four conditions: (1) A separate equipment grounding conductor is provided at each building and enclosure; (2) Separate neutrals are required only where phase-to-neutral loads are supplied; (3) The neutral shall be insulated and isolated from earth, except at one location; (4) An equipment grounding conductor shall be run with the phase conductors.

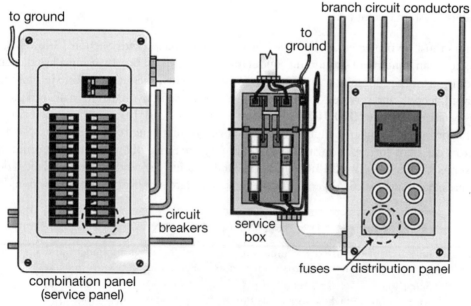

FIGURE 2.1 Overcurrent protection.

Wiring Requirements and Protection

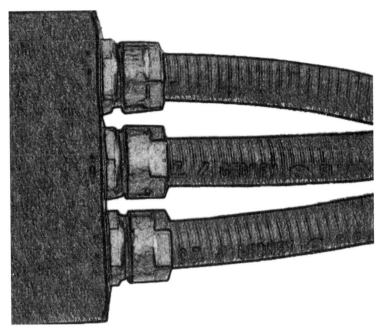

FIGURE 2.2 Flexible conduit.

> **! Code update 250.30(A)(4)**
> A common grounding electrode conductor may connect transformers to a grounding electrode. The minimum size of the common grounding electrode conductor is now 3/0 AWG copper or 250 kcmil Aluminum.

Surge arrestors, which are the protective devices used to limit surge voltage by either discharging or bypassing the surge current, are covered in Article 280. Article 285 covers transient voltage surge suppressors (TVSS). Section **[285.2]** lists the three conditions under which TVSS devices shall not be installed. They are:

1. Circuits which are in excess of 600 volts
2. Any underground systems, impedance grounded systems, and corner grounded delta that are not specifically listed for use on these systems
3. If the TVSS rating is less than the maximum continuous phase-to-ground power frequency voltage which is available at the point of application

Section **[285.5]** states that a TVSS shall be a listed device, and Article 285 II and III cover the installation and connection requirements for TVSS devices.

So just remember, if a question has to do with branch circuits, feeders, service wiring, overcurrent protection, grounding, bonding, or surge protection, it is covered in Chapter two of the NEC.

Now it's time to take a test on your knowledge of these systems. The questions are designed so that you can judge your knowledge of the materials in Chapter two. The first section of the test is a True or False exercise and the answer pages include brief review notes of the material as well as the code references. The second part is an overall review quiz of the requirements for installations and materials throughout the Chapter two of the NEC. Let's get started.

QUIZ 1—PRINCIPLES-ARTICLE 200

1. When electrical current is given multiple conductive paths on which to flow, current will always take the path of least resistance.

   a. True   b. False

2. It is important to ground metal parts to a suitable grounding electrode, so that in the event of a ground fault, dangerous ground-fault current will be shunted into the earth, away from persons, protecting them against electric shock.

   a. True   b. False

3. The grounding conductor for a supplementary grounding electrode (for example, a ground rod for a machine tool) must have the capacity to conduct safely any fault current likely to be imposed on it. This is accomplished by sizing the conductor in accordance with Table 250.66 or Table 250.122, depending on the conditions.

   a. True   b. False

4. Equipment must be grounded so that sufficient fault current will flow through the circuit-protection device to quickly open and clear the ground fault. For example, a 20A circuit breaker will trip and de-energize a 120V ground fault to a metal pole that is grounded to a 25 ohm ground rod.

   a. True   b. False

5. Electrical equipment must be grounded to ensure that dangerous voltage on metal parts resulting from a ground fault can be reduced to a safe value.

   a. True   b. False

6. Metal traffic signal poles and handhole covers must be grounded to a suitable grounding electrode to ensure that dangerous voltage on metal parts resulting from a ground fault can be reduced to a safe value.

   a. True   b. False

7. Grounding of metal manhole covers to a suitable grounding electrode ensures that dangerous voltage on metal parts resulting from a ground fault can be reduced to a safe value.

   a. True   b. False

8. Service equipment must be grounded to a grounding electrode to ensure that dangerous voltage on metal parts, caused by a ground fault, can be removed or reduced to a safe value.

   a. True   b. False

Wiring Requirements and Protection

9. Service equipment is grounded to a grounding electrode to ensure that metal parts subject to a ground fault remain at the same potential as the earth.

   a. True   b. False

10. Grounding of service equipment to a grounding electrode is necessary to stabilize the system voltage.

    a. True   b. False

11. The grounding of service equipment ensures that all metal parts of equipment with which personnel can come into contact are always at or near zero (0) volts with respect to ground (earth).

    a. True   b. False

12. The metal parts of separately derived systems are grounded to ensure that the voltage, as measured between the metal parts of the electrical installation and the earth, remains at the same potential during a ground fault.

    a. True   b. False

13. Separately derived systems must be grounded to a grounding electrode to ensure that dangerous voltage on metal parts, caused by a ground fault, can be removed or reduced to a safe value.

    a. True   b. False

14. An ungrounded system gets its name from the fact that both the separately derived system and the metal case of the separately derived system are isolated from the ground (earth).

    a. True   b. False

15. Failure to ground the metal case of a transformer to a grounding electrode can result in a dangerous difference of potential between the metal parts of different separately derived systems.

    a. True   b. False

16. The metal case of generators is grounded to a suitable grounding electrode to ensure that dangerous voltage on metal parts, caused by a ground fault, can be reduced to a safe value.

    a. True   b. False

17. Building disconnection means that a remote building supplied by a feeder must be grounded to a grounding electrode to ensure that dangerous voltage on metal parts, caused by a ground fault, can be removed or reduced to a safe value.

    a. True   b. False

18. The metal disconnecting means that a remote building, supplied by a feeder with an equipment grounding conductor, is not required to be grounded to a grounding electrode.

    a. True   b. False

19. Outdoor metal light poles must be grounded to a suitable grounding electrode to ensure that dangerous voltage on metal parts, caused by a ground fault, can be reduced to a safe value.

    a. True   b. False

20. Grounding metal light poles to a grounding electrode helps in reducing damage from a direct lightning strike to the luminaires on the metal light pole.

    a. True   b. False

21. Grounding metal light poles to a grounding electrode helps in preventing damage to building wiring and equipment from a lightning strike to one of the metal light poles.

    a. True   b. False

22. Grounding metal light poles to a grounding electrode is necessary to prevent lightning damage to the concrete pole base.

    a. True   b. False

23. Studies have shown that a low-resistive grounding system improves power quality for sensitive electronic equipment.

    a. True   b. False

24. Single-point grounding improves equipment performance by preventing ground-loop currents.

    a. True   b. False

25. Studies have shown that grounding sensitive electronic equipment to an isolated counter-poise ground improves equipment performance.

    a. True   b. False

26. If an electrical system is properly installed and operating normally, there should be no potential (voltage) difference between the neutral terminal and the ground terminal at a receptacle.

    a. True   b. False

27. Grounding premises wiring to a low-resistive grounding grid can help reduce stray voltage or neutral-to-earth voltage on metal parts.

    a. True   b. False

Wiring Requirements and Protection 31

28. A low-resistive earth ground is necessary for the proper operation of transient voltage surge suppressors (TVSSs).

  a. True **b.** False

29. A 115V hair dryer plugged into a GFCI protected receptacle will always trip if it's immersed in water.

  a. True **b.** False

30. Where a lightning protection system is installed, it must be grounded to an independent grounding electrode without any electrical connections to the building electrical system.

  a. True **b.** False

QUIZ 2—REVIEW-ARTICLE 200

1. A single-family dwelling has three bathrooms each with the following: a lighting fixture, a fan, and one receptacle outlet. In one of the bathrooms, the lighting fixture, fan, and receptacle outlet are installed on a dedicated 20 ampere circuit. For this dwelling, the minimum number of 20 ampere circuits required to serve the bathrooms is which of the following:

  a. Two **b.** Three
  c. Four **d.** Five

2. One building is supplied power from another building on a non-industrial property with a single owner. The underground feeder is protected by a 100 ampere circuit breaker in the first building. Qualified persons are not always available to service the installation; therefore, the disconnecting means for the second building must be which of the following:

  a. Located inside the second building, and is not required to be located near the point where the conductors enter the building
  b. At the closest practical point where the conductors enter the building, and may be located either inside or outside of the building
  c. The circuit breaker in the first building
  d. Located on the outside of the building, near the point where the conductors enter the building

3. In a single-phase, 3-wire electrical system, the middle conductor must be which of the following:

  a. Hot **b.** Grounded
  c. Ungrounded **d.** Out-of-phase

4. In a 3-wire, single phase electrical system, the nominal voltage must be 120 volts between the ungrounded conductor and the neutral, and the volts between the two ungrounded conductors must be which of the following:

   a. 120 volts
   b. 240 volts
   c. 288 volts
   d. 600 volts

5. In a single-phase, 3-wire electrical system, the hot conductors are referred to as which of the following:

   a. Neutral conductors
   b. Grounded conductors
   c. Nominal conductors
   d. Ungrounded conductors

6. The current flowing through the neutral of a 120/240 volt 3-wire, single-phase electrical system is calculated as which of the following?

   a. The difference between the current of the two ungrounded conductors
   b. The sum of the current flowing on the two ungrounded conductors
   c. The current of the first ungrounded conductor divided by the current of the second ungrounded conductor
   d. 240 volts divided by 120 volts

7. The exposed non-current carrying metal parts of a hand-held cord-and-plug drill must be grounded in which of the following scenarios:

   a. The power source is greater than 150 volts to ground
   b. The drill is for residential use
   c. The drill is being used in a hazardous location
   d. All of the above

8. If a rod electrode is required for grounding purposes and a layer of rock restricts the rod from being driven into the ground, which of the following alternate methods of installation shall be used:

   a. Connect to the nearest steel section of the building
   b. Connect to the metal water main
   c. Bury the rod in a trench which is a minimum of 2 1/2 feet deep
   d. Bury the rod in steel conduit with a minimum of 18 inches of cover

9. The bonding jumper between a supplemental electrode and service equipment shall not be required to be larger then which of the following:

   a. #12
   b. #10
   c. #8
   d. #6

Wiring Requirements and Protection

10. The maximum overcurrent device size required to protect a 5-foot length of 1-inch liquid tight flexible metal conduit with no equipment grounding conductors and ground listed fitting terminations is which of the following:

    a. 30 amperes
    b. 40 amperes
    c. 60 amperes
    d. 120 amperes

11. Terminals connected to a grounded conductor shall be identified in which of the following ways:

    a. Identification shall be substantially white in color
    b. Connection must use a terminal screw that is not readily removable and is green in color
    c. Identification must include an engraved metal tag
    d. None of the above

12. The receptacle example listed below which may be connected to a small appliance branch circuit is which of the following:

    a. Garage ceiling receptacle for an automatic garage door opener
    b. Any receptacle which is within 20 feet of the kitchen
    c. An electric clock plugged in at the dining room
    d. An electric hair dryer

13. Open conductors which are not service entrance cables shall not be installed less than which of the following:

    a. 3 feet from grade level
    b. 6 feet below grade level
    c. 8 feet below grade level
    d. 10 feet from grade level

14. The service disconnection means in a building shall not have more than how many switches or circuit breakers:

    a. 6
    b. 8
    c. 10
    d. 20

15. The total number of underground conductors for an outside lighting circuit on a single common neutral conductor is which of the following:

    a. 6
    b. 8
    c. There is no limit specified
    d. Underground conductors specified in this example are prohibited

16. If an individual 30 amp branch circuit feeds a single non-motorized equipment receptacle, then the receptacle amperage must be which of the following:

    a. 20
    b. 30
    c. 40
    d. None of the above

17. Conductors run in parallel raceways must have equipment grounding conductors which are which of the following:

    a. Run in individual raceways
    b. Supported every 6 inches
    c. Run in parallel in each raceway
    d. Protected from excessive temperature

18. A surge arrestor of less than 1000 volts nominal shall have a ground connecting conductor which is no smaller than which of the following sizes:

    a. #8
    b. #10
    c. #12
    d. #14

19. When there is more than one nominal voltage system in the building, ungrounded feeder conductors must be which of the following:

    a. Connected through a transfer switch
    b. Separately identified
    c. Run in individual conduits
    d. Identified using the same color

20. Single-point grounding is allowed at the source of a separately derived system when which of the following conditions exists:

    a. A separate equipment grounding conductor is provided at each building and enclosure
    b. The neutral is insulated and isolated from earth, except at one location
    c. An equipment grounding conductor is run with the phase conductors
    d. All of the above

21. When measuring the area of a dwelling in order to determine the lighting load, the outside dimensions are used, which include any porches, garages, or carports.

    a. True
    b. False

22. The continuous load supplied by a branch circuit shall not, under any circumstances, exceed 80% of the conductor ampacity before any derating factors are applied.

    a. True
    b. False

Wiring Requirements and Protection

23. The computed load of a feeder must be 75% or less than the sum of the loads on the branch circuits after any derating factors are applied.

   a. True   **b.** False

24. Under certain conditions, a feeder neutral may be smaller than the ungrounded conductors in a wiring system.

   a. True   **b.** False

25. Branch circuit rating is determined by the size of the overcurrent device.

   a. True   **b.** False

QUIZ 3—TRUE / FALSE

1. The permitted identification of a size #6 AGW or smaller insulated grounded conductor is a continuous white outer finish.

   a. True   **b.** False

2. An equipment grounding conductor that is run with circuit conductors shall not be electrical metallic tubing.

   a. True   **b.** False

3. The permitted identification of a size #6 AGW or larger insulated grounded conductor is a continuous gray outer finish.

   a. True   **b.** False

4. The minimum size for a surge arrestor conductor is #12 AWG copper.

   a. True   **b.** False

5. Circuits over 120 volts but not exceeding 277 volts between conductors shall be permitted to supply power to screw shell type lamp holders.

   a. True   **b.** False

6. A 125 volt single-phase 15 amp bathroom receptacle must have ground-fault interruption protection.

   a. True   **b.** False

7. The approved identification of a size #6 AGW or larger insulated grounded conductor is a continuous white outer finish.

   a. True   b. False

8. Multi-wire branch circuits shall only supply up to two units of utilization equipment.

   a. True   b. False

9. A two-wire AC circuit with two ungrounded conductors is permitted to be tapped from ungrounded conductors of circuits that have a grounded neutral conductor.

   a. True   b. False

10. An arc-fault circuit interrupter device recognizes certain characteristics unique to arcing and de-energizes the circuit when an arc fault is detected.

    a. True   b. False

11. All 120 volt single-phase 20 amp branch circuits in a dwelling unit bedroom shall have a combination type arc-fault circuit interrupter.

    a. True   b. False

12. An electric 8 kW range shall have a minimum branch circuit size of 40 amps.

    a. True   b. False

13. A 20 amp receptacle connected to a 20 amp branch circuit supplying two or more outlets shall not have a total cord-and-plug load that exceeds 24 amps.

    a. True   b. False

14. In a separately derived system, the TVSS shall be connected to the load side of the first overcurrent device.

    a. True   b. False

15. A TVSS device shall not be installed on an impedance grounded system.

    a. True   b. False

16. A surge arrestor shall not be installed on an impedance grounded system.

    a. True   b. False

17. A 125 volt single-phase 20 amp garage receptacle must have ground-fault interruption protection.

    a. True   b. False

Wiring Requirements and Protection

18. A type "S" fuse classified at 16 amps is permitted to be interchangeable with a lower amp classification.

    a. True   **b.** False

19. An impedance grounded neutral system is permitted if ground detectors are installed on the system.

    a. True   **b.** False

20. Where more than one building exists on the same property under single management, additional feeders or branch circuits are permitted to supply optional standby systems.

    a. True   **b.** False

21. Circuits that do not exceed 120 volts between conductors shall only be permitted to supply power to cord-and-plug devices.

    a. True   **b.** False

22. A grounded circuit conductor may be used to ground non current-carrying raceways if located on the supply side of an ac service-disconnection means.

    a. True   **b.** False

23. The minimum size of an equipment grounding conductor used to ground equipment with an automatic overcurrent device rated at 40 amps in the circuit ahead of the equipment piece is #8 AWG copper.

    a. True   **b.** False

24. Barring any exceptions, equipment grounding conductors that are larger than #6 AWG must be permanently identified at each end and at every point where the conductor is accessible.

    a. True   **b.** False

25. Circuits in excess of 600 volts nominal between conductors shall only be permitted to supply equipment located in areas that only a qualified person can access.

    a. True   **b.** False

26. The approved identification of a size #6 AGW or smaller insulated grounded conductor is three continuous white stripes down the length of a green insulated outer finish.

    a. True   **b.** False

27. A 125 volt single-phase 20 amp kitchen counter receptacle must have ground-fault interruption protection.

    a. True   **b.** False

28. A grounding ring must encircle a building and be in direct contact with the earth.
   a. True   **b.** False

29. Metal underground gas pipe shall be permitted to be used as a grounding electrode.
   a. True   **b.** False

30. The circuit breakers used for overcurrent protection of 3-phase circuits must have a minimum of 6 overcurrent relay elements.
   a. True   **b.** False

31. A fuse must be connected on the load side of an overcurrent relay element.
   a. True   **b.** False

32. When there is more than one nominal voltage system in the building, ungrounded feeder conductors may be separately identified by using marking tape.
   a. True   **b.** False

33. Thermal relays are permitted to be used to protect motor-branch circuits.
   a. True   **b.** False

34. An adjustable-trip circuit breaker must be located in a manner that provides restricted access.
   a. True   **b.** False

35. A single-point grounded neutral system may be supplied from a separately-derived system.
   a. True   **b.** False

36. Series rating is permitted for motors connected on the load side of an overcurrent device.
   a. True   **b.** False

37. A common grounding electrode conductor in multiple separately derived systems shall not be smaller than 3/0 AWG copper.
   a. True   **b.** False

38. A feeder overcurrent device that is not readily available shall be installed with the branch circuit overcurrent devices on the load side.
   a. True   **b.** False

Wiring Requirements and Protection 39

39. In a commercial building with a 480/277 service and loads operating at 208/120 volts, all of the ungrounded conductors for both systems would need to be separately identified.

    a. True   **b.** False

40. The minimum size of the common grounding electrode conductor that connects a transformer to a grounding electrode is now 2/0 AWG copper.

    a. True   **b.** False

41. Fixed lighting units in a dwelling may be supplied by a 30 amp branch circuit.

    a. True   **b.** False

42. Any one piece of cord-and-plug-connected utilization equipment shall not have a rating that exceeds 75% of the branch circuit ampere rating.

    a. True   **b.** False

43. Unless an equipment grounding conductor is bare, it must always be identified with a solid green insulated covering.

    a. True   **b.** False

44. Ground fault interrupter protection is required for a receptacle over the sink in a bathroom.

    a. True   **b.** False

45. A wall switch-controlled lighting outlet in a single-family dwelling kitchen shall be permitted to be a receptacle outlet instead of a lighting outlet.

    a. True   **b.** False

46. Branch circuit rating is determined by the size of the overcurrent protective device.

    a. True   **b.** False

47. Branch circuits rated in excess of 20 amps require the use of heavy-duty type lamp holders.

    a. True   **b.** False

48. Three kitchen unit loads in a restaurant have a demand factor of 90%.

    a. True   **b.** False

49. A school with electric space heating or air conditioning shall not have the feeder or service load calculated by using the optional calculation method.

   a. True
   b. False

50. The maximum voltage drop for combined branch circuits and feeders shall not exceed 3% of the circuit voltage.

   a. True
   b. False

QUIZ 4—MULTIPLE CHOICE

1. The permitted identification of a size #6 AGW or smaller insulated grounded conductor is which of the following:

   a. A continuous white outer finish
   b. Three yellow stripes down the length of a green insulated conductor
   c. One blue stripe on a gray insulated conductor
   d. All of the above

2. Branch circuits shall not be derived from autotransformers unless which of the following exist:

   a. The grounded conductor is protected from access by unauthorized personnel.
   b. The autotransformer operates at less than 600 volts nominal.
   c. The circuit supplied has a grounded conductor that is electrically connected to a grounded conductor from the system that supplies the autotransformer.
   d. All of the above

3. A two-wire AC circuit with two ungrounded conductors is permitted to be tapped from ungrounded conductors of circuits that have which of the following:

   a. Two switching devices
   b. An automatic multi-pole switch
   c. A grounded neutral conductor
   d. None of the above

Wiring Requirements and Protection

4. The minimum size of an equipment grounding conductor used to ground equipment with an automatic overcurrent device rated at 20 amps in the circuit ahead of the piece of equipment is which of the following:

   a. #14 AWG copper   **b.** #12 AWG copper

   c. #10 AWG copper   **d.** #8 AWG copper

5. The general lighting load for a barber shop is which of the following:

   a. 3 volt-amperes per square foot

   b. 2 volt-amperes per square foot

   c. 33 volt-amperes per square foot

   d. 1 volt-amperes per square foot

6. Electrical systems that are grounded must meet which of the following requirements:

   a. Be connected to the earth in a way that limits the voltage caused by line surges

   b. Be connected to the earth in a manner that will stabilize the voltage to the earth during normal operation

   c. Both of the above

   d. Either of the above

7. An electric 9 kW range shall have a minimum branch circuit size of which of the following:

   a. 15 amps   **b.** 20 amps

   c. 30 amps   **d.** 40 amps

8. In a separately derived system, the TVSS shall be connected in which of the following manners:

   a. To each ungrounded connector

   b. Outdoors, in an easily accessible location

   c. To the corner grounded delta

   d. To the load side of the first overcurrent device

9. A TVSS device shall not be installed under which of the following conditions:

   a. On an impedance grounded system

   b. If the TVSS is less than the maximum continuous phase-to-phase power frequency that exists at the point of application

   c. On circuits in excess of 600 volts

   d. All of the above

10. The wire size of a grounding ring must be no smaller than which of the following:
    a. #2 AWG bare copper
    b. #6 AWG copper
    c. #4 AWG aluminum
    d. None of the above

11. A common grounding electrode conductor in multiple separately derived systems shall not be smaller than which of the following:
    a. The size of the tap connector
    b. 3/0 AWG copper
    c. 13% of the derived phase conductor
    d. All of the above

12. A 125 volt single-phase kitchen counter receptacle must meet which of the following requirements:
    a. Have ground-fault interruption protection
    b. Be a maximum of 20 amps
    c. Be a minimum of 15 amps
    d. All of the above

13. A surge arrestor may be installed under which of the following conditions:
    a. On an impedance grounded system
    b. On a service less than 1000 volts with a grounding electrode for the service
    c. Either of the above
    d. None of the above

14. The demand factor for the service and feeder conductors for a new all-electric restaurant with a total connected load of 240 kVA is which of the following:
    a. 80%
    b. 100%
    c. 50% of the amount over 200 kVA + 172.5
    d. 10% of the amount over 200 kVA + 160.0

15. An impedance grounded neutral system is permitted to be installed if which of the following conditions is met:
    a. If installed to serve a line-to-neutral
    b. If installed outside in an easily accessible location
    c. Ground detectors are installed on the system
    d. None of the above

Wiring Requirements and Protection 43

16. Multi-wire branch circuits are permitted to supply which of the following:

   a. Only line-to-line neutral loads

   b. Only one piece of utilization equipment

   c. Both of the above

   d. None of the above

17. The minimum size of an equipment grounding conductor used to ground equipment with an automatic overcurrent device rated at 40 amps in the circuit ahead of the piece of equipment is which of the following:

   a. #8 AWG copper   **b.** #8 AWG aluminum

   c. #10 AWG aluminum   **d.** #10 AWG cooper

18. The size of the sole connection of a grounding electrode conductor connected to a concrete-encased electrode shall not be required to be which of the following:

   a. Larger than #4 AWG copper

   b. Connected to an electrode encased by at least 2 inches of concrete

   c. Both of the above

   d. None of the above

19. For a 30 amp receptacle connected to a 30 amp branch circuit supplying two or more outlets, the total cord-and-plug load may not exceed which of the following sizes:

   a. 24 amps   **b.** 16 amps

   c. 15 amps   **d.** 12 amps

20. An equipment grounding conductor that is run with circuit conductors is permitted to be which of the following:

   a. A solid copper busbar

   b. Any flexible metal conduit

   c. Any liquid tight conduit

   d. All of the above

21. Circuits over 120 volts but not exceeding 277 volts between conductors shall be permitted to supply power to which of the following:

   a. The auxiliary equipment of electric-discharge lamps mounted in permanently installed fixtures

   b. Luminaries equipped with mogul-base screw shell lamp holders

   c. Screw shell type lamp holders

   d. None of the above

22. When calculating the demand factor of a multifamily dwelling connected load, which of the following shall be included:

    a. The smaller of the air-conditioning load or the space-heating load

    b. 3 volt-amperes per square foot of dwelling area for general lighting and general use receptacles

    c. 1500 volt-amperes for each 3-wire 20 amp small appliance branch circuit

    d. All of the above

23. A grounding ring shall encircle a building and consist of bare copper in a length of at least which of the following:

    a. 5 feet
    b. 10 feet
    c. 15 feet
    d. 20 feet

24. A type "S" fuse shall be which of the following:

    a. A plug type fuse

    b. Classified at not over 125 volts

    c. Classified at 0 to 15 amperes

    d. All of the above

25. The demand factor for a 10-unit multifamily dwelling is which of the following:

    a. 50 percent
    b. 45 percent
    c. 43 percent
    d. 39 percent

26. The lighting load demand factor for a hospital shall be calculated as which of the following:

    a. 35%

    b. 40% of the first 50,000 volt-amperes

    c. 30% of the amount over 50,000 volt-amperes

    d. None of the above

27. The minimum size copper circuit wire for a 20 amp branch circuit conductor is which of the following:

    a. 8
    b. 10
    c. 12
    d. 14

Wiring Requirements and Protection

28. Restricted-access, adjustable-trip circuit breakers must meet which of the following requirements:
  a. Have a removable and sealable cover over the adjusting means
  b. Be located behind bolted equipment covers
  c. Be accessible only to a qualified person by means of a locked door
  d. All of the above

29. Each plate electrode used for grounding shall expose which of the following:
  a. A #4 AWG copper conductor
  b. At least 0.06 inches above ground
  c. Not less than 2 square feet of surface-to-exterior soil
  d. None of the above

30. The laundry area in a single-family dwelling unit must have which of the following:
  a. A minimum of one 20 amp and one 220 amp receptacle
  b. At least one receptacle
  c. At least one receptacle installed within 3 feet of the washing machine location
  d. A minimum of two GFCI receptacles

31. If a feeder conductor carries the total load supplied by the service conductors with an ampacity of 50 amps, then which of the following standards must be met:
  a. The feeder ampacity must be greater than the service conductor ampacity
  b. The feeder ampacity must be less than the service conductor ampacity
  c. The feeder ampacity must be 30 amps
  d. None of the above

32. Where more than one building exists on the same property under single management, additional feeders or branch circuits are permitted to supply which of the following:
  a. Optional standby systems
  b. Fire pumps
  c. Parallel power production systems
  d. All of the above

33. The use of 5-wire feeders is which of the following:
  a. Prohibited
  b. Restricted to installations over 600 volts nominal
  c. Requires that the overcurrent device protection be 100% of the continuous load
  d. Permitted to use a common neutral

34. DC power systems located on the premises must include which of the following:
    a. A grounding connection at the power source
    b. A grounding electrode conductor which is at least #10 AWG
    c. A grounding ring
    d. All of the above

35. The non current-carrying metal parts of equipment shall be considered effectively grounded by use of which of the following methods:
    a. If it is secured to the structural metal frame of a building
    b. By use of a separate grounded circuit conductor, running in a separate raceway
    c. Either of the above
    d. None of the above

36. An electrode that is permitted as a grounding means is which of the following:
    a. A metal underground water pipe
    b. A plate electrode
    c. A grounding ring
    d. All of the above

37. The grounded conductor for a single-phase 3-wire AC premises wiring system shall be which of the following:
    a. A grounding electrode
    b. The neutral conductor
    c. The common conductor
    d. Either A or B

38. The circuit breakers used for overcurrent protection of 3-phase circuits must have a minimum of three overcurrent relay elements that meet which of the following requirements:
    a. Are operated from three current transformers
    b. Have a neutral which is regrounded on the load side of the circuit
    c. Have a series rating of 125% of the total circuit load
    d. None of the above

39. A fuse must be connected in a manner that meets which of the following requirements:

   a. On an overcurrent relay element

   b. In series with each ungrounded conductor

   c. On the secondary side of a transformer

   d. Either B or C

40. Each set of conductors that feeds separate loads of a transformer secondary shall be connected in which of the following methods:

   a. With an overcurrent device on the multioutlet line of the branch circuit

   b. In series

   c. Without overcurrent protection at the secondary

   d. None of the above

41. A lightning arrestor is a form of which of the following:

   a. Surge arrestor

   b. Service equipment

   c. Impedance means

   d. Fuse

42. A single-point grounded neutral system may include which of the following:

   a. A grounding electrode

   b. A bonding jumper that connects the equipment grounding conductor to a grounding electrode conductor

   c. A grounding electrode conductor that connects the grounding electrode to the system neutral

   d. All of the above

43. Service equipment electrical continuity shall be ensured by which of the following:

   a. Bonding equipment to the neutral conductor

   b. Use of bonding type bushings

   c. Both of the above

   d. Either of the above

44. If a bonding electrode is a rod, then the portion of the bonding jumper that is the sole connection to the supplement grounding electrode shall be which of the following sizes:

   a. Not larger than #4 copper wire

   b. Not larger than #6 copper wire

   c. Not smaller than #6 copper wire

   d. A bare aluminum wire not less than #8 AWG

45. If the use of multiple grounding connections results in objectionable current, which of the following alterations is permitted:
    a. Change the location of the grounding connections
    b. Discontinue one or more, but not all, of the grounding connections
    c. Either of the above
    d. Both of the above

46. A feeder overcurrent device that is not readily available shall be installed in which of the following manners:
    a. Branch circuit overcurrent devices must be installed on the load side
    b. Branch circuit overcurrent devices shall have a lower ampacity rating than the feeder overcurrent device
    c. Branch circuit overcurrent devices must be installed in a readily accessible location
    d. All of the above

47. Open conductors that are not service entrance cables shall not be installed less than which of the following:
    a. 10 feet from grade level
    b. 8 feet below grade level
    c. 5 feet below grade level
    d. 3 feet from grade level

48. In a 3-wire, single-phase electrical system, the nominal voltage must be 120 volts between the ungrounded conductor and the neutral and the volts between the two ungrounded conductors must be which of the following:
    a. 110 volt
    b. 120 volts
    c. 200 volts
    d. 240 volts

49. Ground-loop currents flow if the neutral-to-ground connections are made in which of the following ways:
    a. On the feed side of separately derived systems
    b. On the load side of service equipment or separately derived systems
    c. In front of the overcurrent protection device
    d. All of the above

50. The metal disconnecting means at a remote building, supplied by a feeder with an equipment grounding conductor, is required to be which of the following:

   a. Connected on the service load side of the feeder

   b. Buried not less than 3 feet under the ground

   c. Grounded to a grounding electrode

   d. None of the above

ANSWERS

QUIZ 1—PRINCIPLES-ARTICLE 200

1. B

Review note: In parallel paths, current divides and flows through each individual parallel path in accordance with Kirchoff's current law. So, when given multiple conductive paths on which to flow, current will take all of the available paths. Yes, it's true that more current will flow through the lower resistive path, as compared to a higher resistive path in a parallel circuit, but that's not what the question is asking.

2. B

Review note: A person touching an energized metal pole, which is only grounded, will experience between 90 and 120 mA of current flow through the body, which is more than sufficient to cause electrocution. Remember, in parallel circuits, current divides and flows through each individual parallel path. The only way to make an installation safe from a ground fault is to bond the electrical equipment to an effective ground-fault current path so that the fault current will be more than sufficient to quickly open the circuit protection device and clear the ground fault [250.2 and 250.4(A)(3)].

3. B

Review note: A supplementary electrode is not required to be sized in accordance with the NEC® [250.54]. During a ground fault, the amount of current flowing through the grounding conductor into the earth and to the power supply depends on the circuit voltage and the earth's resistance. Since the earth's resistance is high, it cannot be used as an effective ground-fault current path [250.4(A)(5)] and so the grounding conductor for a supplementary electrode is not sized in accordance with the NEC® [250.54].

4. B

Review note: A ground fault that uses the earth as the fault return path to the power source is not capable of carrying sufficient current to clear the ground fault [250.4(A)(5)] and would result in dangerous voltage between the metal parts and the earth.

5. B

Review note: Grounding metal parts to the earth does not reduce voltage on metal parts resulting from a ground fault because the earth cannot serve as an effective ground-fault current path [250.5(A)(5)]. For example, if you had a 120 volt circuit and a ground rod resistance of 25 ohms, the current flowing through the grounding conductor into the earth and to the power supply would only be only 4.8 amperes, which is not enough to trip the circuit breaker. The only way to make this installation safe from a ground fault is to bond the electrical equipment to an effective ground-fault current path so that the fault current will be more than sufficient to quickly open the circuit protection device and clear the ground fault, removing the potential hazard of touch voltage [250.2 and 250.4(A)(3)].

6. **B**

    Review note: The only way to make this installation safe from a ground fault is to bond the metal traffic signal poles and handhole covers to an effective ground-fault current path [250.2 and 250.4(A)(3)].

7. **B**

    Review note: The only way to make this installation safe from a ground fault is to isolate the manhole cover from energized parts or to bond the metal parts to an effective ground-fault current path [250.2 and 250.4(A)(3)].

8. **B**

    Review note: Yes, you are finding that this test reinforces in many ways that grounding metal parts to the earth does not remove or reduce voltage on metal parts which could result from a ground fault, and no, the earth cannot serve as an effective ground-fault current path based on Section [250.5(A)(5)].

9. **B**

    Review note: Section [250.2, 250.4(A)(3) and 250.24(C)].

10. **B**

    Review note: The earth plays no part in stabilizing system voltage. System voltage is stabilized by the grounding of the utility secondary winding [250.4(A)(1)].

11. **B**

    Review note: The earth doesn't establish or maintain a zero-difference of potential between metal parts of electrical equipment and the earth during a ground fault.

12. **B**

13. **B**

    Review note: The only way to make this installation safe from a ground fault is to bond the metal parts of the separately derived system by using a system bonding jumper so that the ground fault current will quickly open the circuit protection device and clear the ground fault [250.2, 250.4(A)(3), and 250.4(A)(3)].

14. **B**

    Review note: The NEC® requires the metal case of ungrounded separately derived systems to be grounded to a grounding electrode [250.30(B)(1)].

15. **B**

    Review note: Grounding the metal case of a transformer to a grounding electrode is not necessary to reduce the difference of potential between the metal parts of different separately derived systems. This is because there is no difference between the metal parts of the separately derived systems. All metal parts of electrical installations are required to be bonded to an effective ground-fault current path [250.4(A)(3)]. The NEC® requires that the metal case of all separately derived systems be grounded to a suitable grounding electrode [250.30(A)(3) and (7)].

16. **B**

    Review note: *Grounding* metal parts to the earth does not remove or reduce voltage on metal parts resulting from a ground fault because the earth cannot serve as an *effective ground-fault current path* [250.5(A)(5)]. The only way to make this installation safe from a ground fault is to bond the metal case of the generator to an *effective ground-fault current path* so that the fault current will be more than sufficient to quickly open the circuit protection device, thereby clearing the *ground fault* and removing dangerous touch voltage [250.2, 250.4(A)(3) and 250.30(A)(1)].

17. **B**

    Review note: [250.2, 250.4(A)(3) and 250.32(B)].

18. **B**

    Review note: Grounding a remote building disconnection means to the earth will reduce the voltage on the metal parts from lightning and reduce the likelihood of fire. The equipment grounding conductor supplies the required low-impedance path to the source, which is necessary to clear a ground fault. Equipment grounding conductors are not intended to serve as a path for lightning to the earth.

19. **B**

    Review note: The only way to make this installation safe is to bond the metal light pole to an effective ground-fault current path so that the fault current will quickly open the circuit protection device [250.2 and 250.4(A)(3)].

20. **B**

    Review note: Grounding a metal light pole to the earth does nothing to prevent damage to interior wiring and equipment of a building from lightning. Interior wiring and equipment can be protected from lightning-induced voltage transients on the circuit conductors by the use of properly designed TVSS protection devices.

21. **B**

    Review note: If lightning strikes the pole, the luminaire on the pole will be ruined. You can't prevent it.

22. **B**

    Review note: This is an old wives tale. Lightning does not crack the concrete of a concrete-encased grounding electrode.

23. **B**

    Review note: If lightning strikes the pole, the earth isn't going to improve power quality.

24. **B**

    Review note: Grounding sensitive electrical equipment to the same electrode won't prevent or reduce ground-loop currents. Ground-loop currents flow if the neutral-to-ground connections are made on the load side of service equipment or separately derived systems in violation of [250.142].

25. **B**

    Review note: Grounding equipment to an isolated grounding electrode can cause equipment damage when lightning current produces a potential difference between the counter-poise ground and the structure ground.

26. **B**

    Review note: There will always be voltage between the neutral and ground terminals at a receptacle. For example: the NEC® recommends a maximum voltage drop of 3% for the feeder under a load conduction, which works out to be 3.6V for a 120V circuit. Under this condition, the voltage (feeder neutral voltage drop) as measured between the receptacles' neutral and ground terminals would be 1.8V if no current flows through the branch circuit supplying the receptacle. Naturally, if the branch circuit is loaded, the voltage between the neutral and ground terminal would be greater than 1.8V. A study by the Electrical

Power Research Institute (EPRI) demonstrated that elevated neutral-to-ground voltage has no effect on equipment performance.

27. B

Review note: Grounding metal parts to the earth does not reduce neutral-to-earth voltage (NEV). Bonding metal parts together reduces the difference of potential between the metal parts. Stray voltage or NEV can come from the electric utility's distribution system or the building's electric system.

28. B

Review note: TVSS protection devices protect electrical equipment by shunting high-frequency impulse currents away from the load and back to the source via the circuit conductors, not via the earth.

29. B

Review note: If water is contained in a nonmetallic sink where there is no conductive path to the power supply, the GFCI protection device will not trip and the water will be energized.

30. B

Review note: Lightning protection systems are required to be bonded to the building or grounding electrode system [250.106].

QUIZ 2—REVIEW-ARTICLE 200

1. **A** Section [210.11(C)(3)].

> ► **Test** tip
> One circuit is required for the bathroom where the light and fan are combined with the receptacle, and one circuit is required for the receptacle in the other two bathrooms.

2. **B** Section [225.32]
3. **B** Section [250.26]
4. **B** Section [250]
5. **D** Section [250]
6. **A** Section [250]
7. **D** Section [250.114 (1)]
8. **C** Section [250.53 (G)]
9. **D** Section [250.53(E)]
   (NOTE: the question asks for which answer is NOT required.)
10. **C** Section [250.118(6)(c)]
11. **A** Section [200.9]
12. **C** Section [210.52(B)(2)]
13. **D** Section [225.18]

> ► **Test** tip
> This question is full of negatives. First of all, the conductors are NOT service entrance cables and shall NOT be installed less than 10 feet from grade level. Another way to read the question would be, "What is the minimum depth at which non service-entrance conductors shall be installed?"

14. **A** Section [230.71(A)]
15. **C** Section [225.7(B)]
16. **B** Section [210.21(B)]
17. **C** Section [250.122(F)]
18. **D** Section [280.21]
19. **B** Section [215.12]
20. **D** Section [250.184(B)]
21. **B** Section [220.12]
22. **B** Section [210.19(A)(1)] Exception
23. **B** Section [220.40]
24. **A** Section [220.61]
25. **A** Section [210.3]

QUIZ 3—TRUE / FALSE

1. A	14. A	27. A	39. A
2. B	15. A	28. A	40. B
3. A	16. A	29. B	41. B
4. B	17. A	30. B	42. B
5. B	18. B	31. B	43. B
6. A	19. A	32. A	44. A
7. A	20. A	33. A	45. A
8. B	21. B	34. A	46. A
9. A	22. A	35. B	47. A
10. A	23. B	36. A	48. A
11. A	24. A	37. A	49. B
12. B	25. A	38. A	50. B
13. B	26. B		

QUIZ 4—MULTIPLE CHOICE

1. A	14. C	27. C	39. A
2. C	15. C	28. D	40. C
3. A	16. C	29. C	41. A
4. B	17. D	30. B	42. D
5. A	18. A	31. B	43. B
6. C	19. A	32. D	44. B
7. D	20. A	33. D	45. D
8. D	21. B	34. A	46. D
9. D	22. B	35. D	47. A
10. A	23. D	36. D	48. D
11. B	24. D	37. B	49. B
12. D	25. C	38. A	50. C
13. B	26. B		

Chapter 3
WIRING METHODS AND MATERIALS

Electrician's Exam Study Guide

If there is one section of the code that you absolutely need to know in order to perform electrical work and pass the exam, it is Chapter three. The articles in this chapter cover everything from conductors, junction boxes, cable assemblies, service entrance cable and branch circuits to metal and PVC conduits, busways, cablebus and raceways. Chapter two may have introduced you to all of the ingredients you need for standard installations, but Chapter three explains how to connect and install them all together in a safe manner. There are over 60 tables packed full of temperature ratings, bending-space and box size requirements, ampacity and conduit sizing, and allowable fill areas. As you study Chapter three, pay attention to whether the requirements for various installations are listed in *maximum* or *minimum* terms. For example, Table **[312.6(A)]** illustrates the minimum wire-bending space at terminals and the minimum width of wiring gutters, while Table **[314.16(A)]** lists the maximum number of conductors allowed in specific sizes of metal boxes.

> ▶ **Test** tip
>
> When reading an exam question which describes **minimum** allowable installations, you will find key words and phrases such as: at least, no less than, and smallest allowable.
>
> When reading exam questions which are describing **maximum** allowable requirements, look for key words or phrases that include: no greater than, most, greatest, highest, and shall not exceed.

Beginning with Article 300, some of the information is basic. In general, this first section describes wiring methods and ways to protect wiring from damage. Table **[300.19(A)]** lists various sizes and types of wire and the distance, in feet, in which they must be supported. Table **[300.5]** outlines the minimum depth of burial coverage required for various types of installations in the range of 0 to 600 volts nominal. For example, rigid metal conduit that runs under a single-family dwelling driveway must be buried at least 18 inches underground. The same rigid metal conduit installed under a street or highway must be at least 24 inches deep. Larger circuit sizes, from over 600 volts nominal to 40 kV, and the required minimum burial requirements are list in Table **[300.50]**.

> **! Code** update SECTION [300.34]
>
> The required bending radius is now 12 times the *overall* diameter of the shielded or lead-covered conductor.

Article 310 covers general wiring, including designations, markings, insulation, strength, ampacity ratings, and approved uses. An extensive listing of conductor applications and insulation ratings are given in Table **[310.13]**. For example, underground feeder cables and branch circuits, type UF, that are 14 – 10 AWG and are installed in a moisture-resistant manner shall have an operating temperature that does not exceed (maximum) 140 degrees F.

> **! Code** update SECTION [310.4]
>
> When paralleled conductors are installed in separate raceways, the same number of conductors must be used in each raceway or cable.

Wiring Methods and Materials

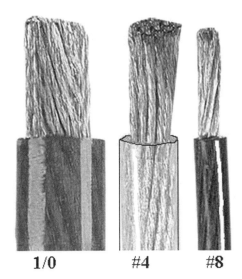

FIGURE 3.1 AWG.

Conductor requirements are described in Article 312. This includes conductors entering and leaving enclosures such as metal cabinets, cutout boxes, and meter socket enclosures, and the regulations for enclosure construction.

Article 314 describes provisions covering boxes, such as device, pull, and junction boxes and outlets, as well as conduits and fittings. Boxes containing 6 AWG and smaller conductors have a maximum cubic-inch capacity, and boxes containing 4 AWG and larger conductors have a minimum size requirement based on the size and number of raceway entries. Also covered in Article 314 are manholes, pull boxes, and junction boxes for systems larger than 600 volts nominal. Table **[314.16(A)]** illustrates the maximum number of conductors allowed by box size.

FIGURE 3.2 Conduit and manhole.

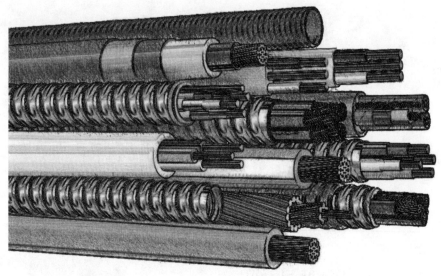

FIGURE 3.3 THHN.

> **! Code** update SECTION **[312.2(A)]**
> A new sentence has been added to **[312.2(A)]**: "For enclosures in wet locations, raceways or cables entering above the level of uninsulated live parts shall use fittings listed for wet locations." This simply means that fittings located above un-insulated parts must prevent water from entering the enclosure

Specifications for the installation of various cable types begin with Article 324 and run through Article 340. For metal clad cabling, type MC, installation and permitted uses are described in Article 330. Unless this cable is specifically identified otherwise or complies with Section **[300.5]** or **[300.50]**, MC cabling is not allowed to be direct buried or encased in concrete. Table **[310.13]** provides a listing of the many types of conductors by trade name, type, and application.

> **! Code** update SECTION **[338.10(B)(4)(b)]**
> Type USE cable is no longer permitted for indoor branch circuits or feeders. The 2002 Code permitted USE cable to be terminated at indoor enclosures if the length of the cable was not more than 6 feet. The 2005 Code now states that USE cable must be protected from physical damage per **[300.5(D)]**.

Conduit specifics for rigid metal conduit begin in Article 344, and flexible metal conduit is covered in Article 348. Rigid metal conduit (RMC) is a threaded metal means of protecting and routing cables and conductors and is generally made of steel or aluminum. The difference in construction with flexible metal conduit (FMC) is that FMC is made of helically wound, formed and interlocked metal stripping. Liquidtight

Wiring Methods and Materials

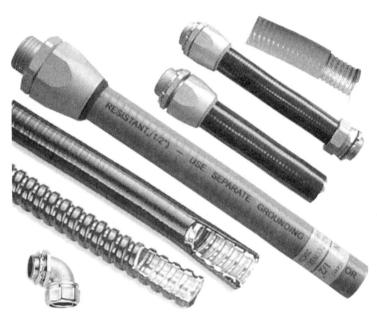

FIGURE 3.4 Liquidtight.

flexible metal conduit, type LFMC, is flexible metal raceway that has a liquidtight, non-metallic, sunlight resistance jacket. Articles 350 lists the installation, size, support, and bend specifications for LFMC.

A busway is a metal enclosure that is grounded and has factory-mounted, bare, or insulated conductors that are usually copper or aluminum rods, bars, or tubes. Busway installations are described in Article 368. Sections **[368.12(A)]** through **[368.12(E)]** describe restrictions for busways and conditions under which they cannot be used. Sheet metal troughs, with hinged or removable covers, that are used to protect electric wires and cables, are called wire raceways and are explained in Article 376. Section **[376.22]** explains the number of conductors allowed in a wire raceway.

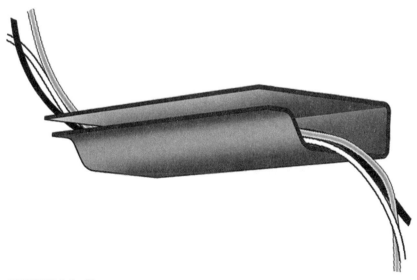

FIGURE 3.5 Raceway.

> **! Code update SECTION [366.58(B)]**
>
> For straight pulls the distance between raceways cannot be less than 8 times the trade size of the largest raceway. For angle pulls, the distance between the raceway and the opposite wall of the box is 6 times the trade size of the largest raceway, plus the sum of all the other raceways on the same wall of the box.

Guidelines for strut-type channel raceways are provided in Article 384 and go through Article 390. Derating factors, which are provided in Section **[310.15(B)(2)(a)],** do not apply to strut-type channel raceway installations if several conditions are met. The calculation method for determining the number of conductors allowed in these raceways is provided in Section **[384.22(3)]**. The number of wires (n) is equal to the channel area in square inches (ca) divided by the wire area (wa).

> **! Code update SECTION [386.70, 388.70]**
>
> The 2005 NEC® requires that signaling circuits be run in different compartments from lighting and power circuits and allows color coding, stamping, and imprinting as approved methods of identifying the separate compartments.

Cable tray installations, including ventilated channels and troughs, ladder, and solid bottom systems are described in Article 392. Table **[392.3(A)]** is a very helpful quick reference because it outlines various types of wiring methods and in which articles they are found. Article 394 covers knob-and-tube wiring and Article 396 and Chapter three of the NEC® concludes with Article 398 on open wiring on insulators.

FIGURE 3.6 Cable tray.

Wiring Methods and Materials

Before you take the following quiz on this chapter, take a few moments to review the various tables throughout Article 300. Next, go over the various definitions of the methods covered in the chapter, so you understand things like busways as described in the NEC. Now, take your time and try to answer correctly as many of the following questions as you can. When you are finished, be sure to review the code references for any of the answers you got wrong, so you can improve your score when you take one of the comprehensive tests later in this book. Immediately following this text is Quiz 1. After that is a Review Test (Quiz 2) on Chapters one through three of the NEC, designed to see how well you can do on combined test questions.

QUIZ I—ARTICLE 300

1. Based on the 2005 NEC, a 4-inch square metal box with a depth of 1 1/2 inches has a volume equal to which of the following:

   a. 16 inches
   b. 21 inches
   c. 32 inches
   d. 33 inches

2. Multiple equipment grounding conductors or equipment bonding jumpers entering a box count as how many total conductors:

   a. One
   b. Two
   c. One per each grounding conductor and one per each bonding jumper
   d. Each bonding jumper counts as one total conductor, but each grounding conductor counts as one per conductor.

3. Conductors must be protected from abrasion only at the exit point of a box, conduit body or fitting:

   a. True
   b. False

4. Where a cable enters a single-gang, non-metallic box through a cable knockout or opening, which of the following must be true:

   a. The cable must extend into the box by 1 1/2 inches.
   b. The cable must be terminated immediately inside of the box.
   c. The cable sheath shall extend into the box by at least 1/4 inch.
   d. The cable shall not be spliced.

5. Boxes in walls constructed of combustible material shall not extend beyond the finished wall surface:

   a. True
   b. False

6. In a straight-pull, multiple raceway, the length must be at least which of the following:

   a. 5 times the trade diameter of the smallest single raceway
   b. 8 times the trade diameter of the largest single raceway
   c. 1 1/2 times the total of all of the raceways combined
   d. Equal to the trade diameter of the largest single raceway

Wiring Methods and Materials

7. The total maximum bend for EMT conduit is which of the following:
- **a.** 90 degrees
- **b.** The total of the two greatest bend angles
- **c.** 180 degrees
- **d.** 360 degrees

8. EMT conduit shall be securely fastened within _____ of a junction box:
- **a.** 6 inches
- **b.** 12 inches
- **c.** 24 inches
- **d.** 36 inches

9. ENT must be secured at which of the following intervals:
- **a.** Each bend
- **b.** Within 1 foot of entering or leaving a raceway
- **c.** Every 3 feet
- **d.** Within 1 foot of entering or leaving an outlet box

10. The maximum trade size for flexible metal conduit is which of the following:
- **a.** 1 1/2 inches
- **b.** 2 inches
- **c.** 3 inches
- **d.** 4 inches

11. A nine foot run of FMC must be secured at which of the following intervals:
- **a.** Each bend
- **b.** Every 4 1/2 feet if a grounding conductor is required
- **c.** Every 3 feet
- **d.** Within 1 foot of entering or leaving an outlet box or electrical cabinet

12. The minimum trade size allowed for RMC is which of the following:
- **a.** 1/4 inch
- **b.** 1/2 inch
- **c.** 1 inch
- **d.** 3 inches

13. Outlet boxes that are mounted in non-combustible ceilings must not be set back more than which of the following:
- **a.** 1/8 inch
- **b.** 1/4 inch
- **c.** 1/2 inch
- **d.** None of the above; outlet boxes must be mounted flush in non-combustible ceilings.

14. Surface metal raceways shall be permitted in which of the following:
    a. Dry locations
    b. Applications less than 300 volts nominal between conductors
    c. In any hazardous classification location
    d. All of the above

15. If an ungrounded #4 or larger conductor enters a raceway in a pull box, it shall meet which of the following requirements:
    a. Be encased in concrete
    b. Be protected by an insulated bushing
    c. Carry less than 600 volts nominal
    d. None of the above

16. In a 200 amp, 120/240 volt single-family dwelling service entrance, with a calculated demand load of 150 amperes, the minimum type THWN aluminum conductor size allowed is which of the following:
    a. 1/0 AWG
    b. 2/0 AWG
    c. 3/0 AWG
    d. 4/0 AWG

17. The minimum rigid metal conduit trade diameter allowed for eight size 10 AWG THWN conductors is which of the following:
    a. 3/8 inch
    b. 1/4 inch
    c. 1/2 inch
    d. 3/4 inch

18. If a nonmetallic single-gang box is installed in a wall, clamping type NM-B cables to the box is not required so long as the cable is supported 8 inches from the box:
    a. True
    b. False

19. When protected solely by enamel, which of the following shall not be installed in outdoor or wet locations:
    a. Ferrous raceways
    b. Fittings
    c. Boxes
    d. All of the above

20. The total number of quarter bends allowed in a single run of rigid nonmetallic conduit shall not exceed which of the following:
    a. 1
    b. 2
    c. 4
    d. 8

Wiring Methods and Materials

QUIZ 2—COMPREHENSIVE TEST-NEC® CHAPTERS 1 THROUGH 3

1. The Authority Having Jurisdiction may grant exception for the installation of conductors and equipment that are inside or outside of a building:

    a. True
    b. False

2. A standard rating for an overcurrent device is which of the following:

    a. 50 amperes
    b. 110 amperes
    c. 275 amperes
    d. 1500 amperes

3. If electrical conductors are run under a parking lot in rigid metal conduit covered by a 4-inch concrete slab, they must be installed at a depth from the finished grade level to the top of the concrete of not less than which of the following depths:

    a. 24 inches
    b. 16 inches
    c. 18 inches
    d. 12 inches

4. If a college owns and maintains its own generators, transmission and distribution lines, the Authority Having Jurisdiction may determine that the National Electrical Safety Code, not the NEC, regulations apply:

    a. True
    b. False

5. A device may do which of the following:

    a. Utilize energy
    b. Carry energy only
    c. Carry or control energy
    d. None of the above

6. If set of 400 amp fuses protect a group of copper feeder conductors with 75 degrees C insulation and terminations with a calculated feeder load of 340 amperes and there are no derating factors, the minimum conductor size allowed for the feeder is which of the following:

    a. 500 kcmil
    b. 485 kcmil
    c. 340 kcmil
    d. 300 kcmil

7. A main bonding jumper is which of the following:

    a. The connection between the grounding conductor and an equipment grounding conductor at a separately derived system
    b. The connection between a grounding conductor and equipment operating at 600 volts nominal or less
    c. The connection between the grounded circuit conductor and the equipment grounding conductor at the service
    d. The connection between the main electrical service and a grounding conductor

8. Under no conditions shall MC cabling be direct buried or encased in concrete:
   a. True   b. False

9. Explanations of arc flash hazards, protection, and calculations are provided in which of the following:
   a. National Industrial Code   b. NFPA® 70
   c. NEC   d. NFPA® 70e

10. The size of conductors installed in a cablebus shall not be smaller than which of the following:
    a. #1/0 AWG   b. #8 AWG
    c. 500 kcmil   d. 250 kcmil

11. Surface metal raceways are permitted in installations where the voltage is 300 volts or more between conductors:
    a. True   b. False

12. When designing an industrial plant which will own and maintain its own generators, distribution lines, and transmission, a certified Electrical Engineer may base the systems on the National Electrical Safety Code instead of the NEC:
    a. True   b. False

13. Where structural members interface, rigid metal conduit shall be allowed to be fastened at which of the following distances from the interface:
    a. 3 feet   b. 5 feet
    c. 4 feet   d. 1 foot

14. Flexible metal conduit and tubing is permitted for equipment grounding if the length in a ground return path does not exceed which of the following:
    a. 3 feet
    b. 4 feet
    c. 6 feet
    d. Flexible metal conduit and tubing is not permitted for this use.

15. Conductors larger than 6 AWG shall not be required to be marked in conduit if there are no splices or unused hubs:
    a. True   b. False

Wiring Methods and Materials

16. The electrical continuity at service equipment and service conductors must be ensured by using which of the following installation techniques:
    a. Threaded couplings or threaded bosses on enclosures are made wrenchtight
    b. Bushings with bonding jumpers are used
    c. Either of the above
    d. Both of the above

17. Lightning protection system ground terminals must be bonded to which of the following:
    a. The building or structure
    b. The building or structure grounding electrode systems
    c. To a common grounding electrode conductor
    d. A surge arrestor

18. Enclosures around electrical equipment enclosures shall be considered accessible if they are controlled by which of the following:
    a. Locks with keys
    b. Combination locks
    c. Keypad locks to which a qualified person has the code in order to unlock the enclosure.
    d. Any of the above

19. A system bonding jumper is which of the following:
    a. The connection between the grounded circuit conductor and the equipment grounding conductor at a separately derived system
    b. A wiring system that connects two or more separately derived systems
    c. The connection between the grounded circuit conductor and the equipment grounding conductor at the service
    d. A wiring system that connects a grounded circuit conductor to equipment

20. Nonmetallic sheathed cable must be secured in place at which of the following intervals:
    a. Every 6 feet
    b. Every 3 feet
    c. Every 48 inches
    d. Every 54 inches

21. Type AC cables are allowed for which of the following uses:
    a. In exposed work
    b. In concealed work
    c. Both A & B
    d. None of the above

22. A ground ring must be in direct contact with the earth at a depth below the ground surface of at least which of the following:
    a. 1 foot
    b. 2 feet
    c. 2 1/2 feet
    d. 3 feet

23. Outdoor electrical installations with live parts operating at 601 volts nominal must have a minimum distance between the equipment and a fenced enclosure of which of the following:
    a. 6 feet
    b. 7 feet
    c. 8 feet
    d. 10 feet

24. Branch circuit conductors of 600 volts nominal or less must have an ampacity not less than the maximum load which is to be served:
    a. True
    b. False

25. Size #14 AWG, type THHN cable is permitted in which of the following applications:
    a. Wet locations
    b. Dry locations
    c. Both of the above
    d. None of the above

26. For a two-family dwelling unit located at grade level, how many outlets are required, assuming installation is no more than 6 1/2 feet above grade level:
    a. One outlet at each exterior door
    b. One outlet at the front of the dwelling unit
    c. A total of two outlets
    d. One outlet at the front and one at the back of the unit

27. A branch circuit that consists of at least two ungrounded conductors with a potential difference between them and an identified grounded conductor which has equal potential between it and each of the ungrounded conductors is considered which of the following:
    a. A split branch circuit
    b. A multi-wire branch circuit
    c. A grounded branch circuit
    d. None of the above

28. A single open conductor which does not serve as a service entrance cable must be installed at which of the following depths:
    a. Not less than 10 feet from grade level
    b. A minimum of 6 feet from grade level
    c. Not more than 10 feet from grade level
    d. Not less than 6 feet from grade level unless encased in concrete

Wiring Methods and Materials

29. A guest suite in a hotel is considered a dwelling unit:
   a. True   **b.** False

30. The soft conversion for 3 feet is which of the following:
   a. 900 millimeters
   b. Less accurate than the hard method of metric conversion
   c. 914.4 millimeters
   d. The only conversion method permitted by code

QUIZ 3—TRUE / FALSE

1. Open wiring on insulators is only permitted for industrial installations of 600 volts nominal or less.
   a. True   **b.** False

2. A nonmetallic extension shall not be used on an existing 15 amp branch circuit.
   a. True   **b.** False

3. A messenger supported wiring system is permitted in hoistways.
   a. True   **b.** False

4. If a 3-inch wide ventilated channel cable tray contains single-conductor cables, then the sum of the diameters of all of the single conductors cannot exceed the inside width of the channel.
   a. True   **b.** False

5. Without exception, EMT shall not be threaded.
   a. True   **b.** False

6. Cable trays must always be corrosion resistant.
   a. True   **b.** False

7. If 2/0 AWG single-conductor cables are installed in a ladder cable tray, then the maximum rung spacing in the cable tray is 9 inches.
   a. True   **b.** False

8. RNC shall be permitted in locations that are subject to severe corrosive influences.
   a. True   b. False

9. Open conductors within 14 feet from the floor are considered to be exposed to physical damage.
   a. True   b. False

10. Single conductors that are installed in a cable tray may be used as equipment grounding conductors.
    a. True   b. False

11. Crossings of two type FCC cable runs shall be permitted.
    a. True   b. False

12. Underfloor raceways shall not be installed under concrete.
    a. True   b. False

13. Type MI cables shall be permitted to be installed in dry, wet, or continuously moist locations.
    a. True   b. False

14. Surface nonmetallic raceways shall be permitted to be used where the voltage is 400 volts between conductors.
    a. True   b. False

15. If the cross-sectional area of a strut-type channel raceway exceeds 4 square inches, then derating factors shall not apply to the conductors installed in the raceway.
    a. True   b. False

16. Unless otherwise listed, RNC shall be securely fastened within 2 feet of a conduit termination.
    a. True   b. False

17. The snap-fit metal cover on a strut-type metallic channel raceway may be used as a means of providing electrical continuity for any receptacles mounted in the cover.
    a. True   b. False

18. The number of conductors allowed in a 1 1/2 inch × 3/4 inch strut-type metallic channel raceways with internal joiners shall be based on a 0.212 square inches wire fill calculation.
    a. True   b. False

Wiring Methods and Materials

19. Splices and taps are not permitted within a nonmetallic wireway under any conditions.
   a. True   **b.** False

20. Metal wireways shall not be permitted for use in exposed or severely corrosive environments.
   a. True   **b.** False

21. The maximum cable fill area for multi-conductor cables rated less than 2000 volts that are installed in a solid channel cable tray with a 3-inch inside width is 2.0 square inches.
   a. True   **b.** False

22. A cablebus is allowed to be used for branch circuits.
   a. True   **b.** False

23. Concealed knob-and-tube wiring is not permitted in the hollow spaces of walls or ceilings.
   a. True   **b.** False

24. The maximum approved length for RMC is 10 feet.
   a. True   **b.** False

25. The minimum size permitted for FMT is 1/2 inches trade size.
   a. True   **b.** False

26. Under no conditions shall the conductors in a cablebus be smaller than 1/0 AWG.
   a. True   **b.** False

27. No more than five #16 AWG conductors shall be permitted in a 3 × 2 × 2 metal device box.
   a. True   **b.** False

28. A busway is a grounded nonmetallic enclosure that contains factory-mounted conductors, which are either bare or insulated, and usually copper or aluminum bars, rods or tubes.
   a. True   **b.** False

29. Without exception, overcurrent protection is required wherever a busway is reduced in ampacity.
   a. True   **b.** False

30. Branches from busways shall be permitted to use Type MI cable.
   a. True   **b.** False

31. Load-break switching devices in a busway shall be interlocked to prevent operation under load.
   a. True   **b.** False

32. The minimum size permitted for rigid nonmetallic conduit is 1/2 inch.
   a. True   **b.** False

33. A busway must have a permanent nameplate which indicates the manufacturer's name and or trademark.
   a. True   **b.** False

34. Intermediate metal conduit is permitted for installations in all types of occupancies and any atmospheric conditions.
   a. True   **b.** False

35. An auxiliary gutter may be used to enclose switches and overcurrent devices.
   a. True   **b.** False

36. Correction factors need not be applied to current-carrying conductors in a sheet metal auxiliary gutter if there are only 25 conductors contained in the gutter.
   a. True   **b.** False

37. Metal auxiliary gutters must be grounded.
   a. True   **b.** False

38. Electrical nonmetallic tubing may be installed concealed in the walls of any building which does not exceed three floors above grade.
   a. True   **b.** False

39. Unless otherwise specified, ENT shall not be used to support luminaries or other equipment.
   a. True   **b.** False

40. A raceway that is circular in cross section, metallic, flexible and liquidtight without a nonmetallic jacket is called Liquidtight Flexible Nonmetallic Conduit type LFNC.
   a. True   **b.** False

41. Electrical metallic tubing shall be permitted for both exposes and concealed work.
   a. True   **b.** False

42. High-density polyethylene conduit shall be installed only in straight runs without bends.
   a. True   **b.** False

43. A vapor seal shall be required in a forced-cooled busway.
   a. True   **b.** False

44. One inch RNC shall be supported every 3 feet.
   a. True   **b.** False

45. A number 6 copper conductor installed in a vertical raceway shall be supported at intervals of 80 feet.
   a. True   **b.** False

46. Liquidtight flexible metal conduit must always be securely fastened every 3 feet.
   a. True   **b.** False

47. FMC is permitted in both exposed and concealed installations.
   a. True   **b.** False

48. Metric designators and trade sizes listed in the NEC® for conduit sizes represent the exact exterior dimensions of each type of conduit.
   a. True   **b.** False

49. For RMC and IMC there shall not be more than 4 quarter bends equal to 360 degrees total between pull points such as conduit bodies and boxes.
   a. True   **b.** False

50. Type FCC cables shall not be installed in a manner that allows polarization of connections.
   a. True   **b.** False

51. Running threads shall be permitted on IMC for the connection of couplings.
   a. True   **b.** False

52. A header shall be attached to a floor duct at no greater than a 45 degree angle.
   a. True   **b.** False

53. Strut-type metallic channel raceways are permitted in exposed dry locations.
    a. True   b. False

54. Underground feeder cable shall not be used as service-entrance cable.
    a. True   b. False

55. A messenger supported wiring system consists of a concealed wiring support system that uses a messenger wire to support insulated conductors.
    a. True   b. False

56. Type AC cable is permitted to connect motors to the motor supply.
    a. True   b. False

57. If 6 current-carrying conductors are installed in a raceway then the maximum load current for each conductor shall be reduced by 70 percent.
    a. True   b. False

58. The edge of holes drilled for cables must be no less than 1 1/4 inches from the nearest edge.
    a. True   b. False

59. Multi-conductor medium voltage cables that run in a cable tray shall be Type MV.
    a. True   b. False

60. Type USE service entrance cable that is identified for underground use in a cable assembly may have a bare copper concentric conductor applied.
    a. True   b. False

61. A cellular concrete floor raceway is allowed to supply ceiling outlets in a commercial garage.
    a. True   b. False

62. The ampacity of type UF cable shall be that of 60 degrees F conductors.
    a. True   b. False

63. FCC cable receptacles and connections shall not be polarized.
    a. True   b. False

Wiring Methods and Materials

64. Two metals of different materials are prohibited from being joined together in order to prevent galvanic action.

    a. True   **b.** False

65. The minimum bending radius for trade size 1 1/4 inches NUCC with conductors is 18 inches.

    a. True   **b.** False

66. Splices in type NUCC shall not be permitted in junction boxes.

    a. True   **b.** False

67. Underground service entrance cable must have a flame-retardant covering.

    a. True   **b.** False

68. Rigid metal conduit is permitted for installations in all types of occupancies and any atmospheric conditions.

    a. True   **b.** False

69. Type NMS cables consist of insulated power or control conductors with signaling, data, and communications conductors enclosed with a metallic jacket.

    a. True   **b.** False

70. If the outer sheath of type MI cable is made of copper then the cable shall be permitted as an equipment grounding means.

    a. True   **b.** False

71. Metal clad cables shall not be used as service cabling.

    a. True   **b.** False

72. Medium voltage cables shall not be installed where exposed to direct sunlight.

    a. True   **b.** False

73. The minimum bend size for type IGS cabling trade size 4 is 45 inches.

    a. True   **b.** False

74. FCC systems are complete wiring systems for branch circuits that are designed to comply with communication installation requirements.

    a. True   **b.** False

75. Alterations to FCC cable systems are prohibited.
    a. True   **b.** False

76. Power and control tray cable, type TC, shall be permitted to be installed in cable trays.
    a. True   **b.** False

77. The ampacity for 1/0 AWG single conductors in an uncovered cable tray shall not exceed 260 volts.
    a. True   **b.** False

78. The temperature rating for FCC cable shall be marked on the cable surface at intervals not to exceed 3 feet.
    a. True   **b.** False

79. Angle connectors and straight fittings are permitted on LFNC direct burial installations.
    a. True   **b.** False

80. Where conduits requiring the use of locknuts or bushings are connected to the side of a box, round boxes shall be used.
    a. True   **b.** False

81. NUCC is permitted to be installed encased in concrete.
    a. True   **b.** False

82. The maximum ampacity of a 1/0 copper conductor is 297 amps.
    a. True   **b.** False

83. Solid dielectric insulated 2000 volt conductors that are permanently installed shall have ozone-resistant insulation and will be shielded.
    a. True   **b.** False

84. Type NM cable is prohibited in single-family dwellings.
    a. True   **b.** False

85. Conductors run in parallel in each phase, polarity, neutral, or grounded circuit conductor may be comprised of various conductor materials.
    a. True   **b.** False

Wiring Methods and Materials

86. A 4 × 4 × 1 1/2 feet metal box can hold a maximum of nine #12 AWG conductors.

    a. True   **b.** False

87. Under no conditions shall open wiring on insulators be installed if they are concealed by the building structure.

    a. True   **b.** False

88. Nonmetallic sheathed cable shall be secured within 12 inches of every box.

    a. True   **b.** False

89. All boxes must have an internal depth of at least 1/2 inch.

    a. True   **b.** False

90. Under all conditions, a busway must be grounded.

    a. True   **b.** False

91. Splices and taps in underfloor raceways shall only be made in junction boxes.

    a. True   **b.** False

92. Metal boxes shall be grounded unless installed in brick or plaster walls or ceiling.

    a. True   **b.** False

93. FMC angel connectors are permitted for both exposed and concealed installations.

    a. True   **b.** False

94. A tapped hole shall be provided in each metal box over 100 cubic inches for the connection of an equipment grounding conductor.

    a. True   **b.** False

95. When more than one calculated ampacity could apply for a specific circuit length, then the highest value shall be used.

    a. True   **b.** False

96. The area in square inches for #8 bare conductors in a raceway is 0.778.

    a. True   **b.** False

97. The maximum distance permitted between 3/4 inch RMC supports is ten feet.
   a. True   **b.** False

98. Integrated gas spacer cabling may be direct buried and used as a grounding conductor.
   a. True   **b.** False

99. A length of conduit which is 24 inches or less is considered a nipple.
   a. True   **b.** False

100. A switch box installed in a tiled wall may be recessed 1/4 inch behind the finished wall.
   a. True   **b.** False

QUIZ 4—MULTIPLE CHOICE

1. Open wiring on insulators is only permitted for which of the following installations:
   a. Industrial installations of 600 volts nominal or less
   b. Agricultural installations of 600 volts nominal or less
   c. Either of the above
   d. None of the above

2. Type FCC cabling systems are approved for installation in which of the following applications:
   a. Residential   **b.** Schools and hospitals
   c. Both of the above   **d.** None of the above

3. If open conductors are exposed to physical damage, they must be installed in which of the following:
   a. Rigid metal conduit   **b.** Rigid nonmetallic conduit
   c. Intermediate metal conduit   **d.** Any of the above

4. A messenger supported wiring system may support insulated conductors by which of the following means:
   a. Factor-assembled aerial cable
   b. A messenger with rings and saddles
   c. Either of the above
   d. None of the above

Wiring Methods and Materials

5. Approved sizes for nonmetallic underground conduit with conductors type NUCC are which of the following:

   a. Minimum size of no less than 3/4 inch trade size and maximum of no larger than size 3

   b. Minimum size of no less than 1/2 inch trade size and maximum of no larger than size 4

   c. Minimum size of metric 21 and maximum size of metric 103

   d. Six foot run lengths

6. Concealed knob-and-tube wiring is not permitted in which of the following applications:

   a. Commercial garages
   b. Motion picture studios
   c. Hazardous classified locations
   d. All of the above

7. Rigid metal conduit shall be permitted in which of the following installations:

   a. Any type of occupancy
   b. Corrosive environments
   c. Wet locations
   d. All of the above

8. The number of 2001 volt cables permitted in a single cable tray shall not exceed which of the following:

   a. 3
   b. 4
   c. 6
   d. None of the above

9. A 6-inch ventilated channel cable tray that contains single-conductor cables must comply with which of the following:

   a. The sum of the diameters of all of the single conductors cannot exceed the inside with of the channel.

   b. The conductors can not exceed 600 volts nominal.

   c. The conductors shall not have a cross-sectional diameter over 600 kcmil.

   d. All of the above

10. HDPE conduit is permitted to be used in which of the following locations:

   a. Areas subject to severe corrosive factors

   b. Hazardous classified locations

   c. Areas with ambient temperatures over 50 degrees C

   d. All of the above

11. The maximum cable fill area in square inches for 3 multi-conductor cables rated less than 2000 volts that are installed in a solid channel cable tray with an inside width of 6 inches is which of the following:

   a. 5.5
   b. 3.2
   c. 2.0
   d. None of the above

12. Multi-conductor power, lighting, and control cables must be installed in a solid bottom cable tray based on which of the following:
    a. If all of the cables are 4/0 AWG the sum of the diameters of all of the cables cannot exceed 90% of the cable tray width.
    b. If all of the cables are 4/0 AWG the cables shall not be installed in a single layer.
    c. If all of the cables are 4/0 AWG the cable fill area shall not exceed 7.0 square inches.
    d. All of the above

13. Medium voltage cable used on power systems rated up to 35,000 volts shall be approved for use in which of the following installations:
    a. In dry or wet locations
    b. In raceways
    c. In messenger-supported wiring
    d. All of the above

14. If 4/0 AWG single-conductor cables are installed in a ladder cable tray, then the maximum rung spacing in the cable tray shall be which of the following:
    a. 18 inches
    b. 9 inches
    c. 4 1/2 inches
    d. 2 inches

15. Single conductors that are used as equipment grounding conductors and that are installed in a cable tray must meet which of the following requirements:
    a. Be insulated, covered or bare
    b. Must be #4 AWG or larger
    c. Either A or B
    d. Both A and B

16. Half-round underfloor raceways which do not exceed 4 inches in width shall be installed in which of the following manners:
    a. With not more than 20 mm of wood above the raceway
    b. With not less than 3/4 inches of wood above the raceway
    c. Either of the above
    d. None of the above

17. Surface nonmetallic raceways shall be permitted for use in which of the following locations:
    a. Dry
    b. Hoistways
    c. Damp
    d. All of the above

Wiring Methods and Materials

18. Strut-type metallic channel raceways are permitted to be used with which of the following applications:
    a. As raceways where the voltage is over 600 volts
    b. As power poles
    c. In damp, exposed locations
    d. All of the above

19. Type MI cable conductors shall be which of the following:
    a. Copper
    b. Nickel-coated copper
    c. Either of the above
    d. None of the above

20. The number of conductors allowed in a 1 1/2 inch × 1 1/2 inch strut-type metallic channel raceways with internal joiners shall be calculated using which of the following methods:
    a. A factor of 0.731 square inches
    b. By dividing the total diameter of the conductors by the channel area in square inches
    c. By using a 25% wire fill value
    d. All of the above

21. Where splices or taps are installed in a metallic wireway, the conductor size including the splices of taps at that point of the wireway shall not fill the wireway in excess which of the following:
    a. 75% of the area at that point
    b. 45% of the area
    c. 25% of the area at the splice or tap point
    d. None of the above

22. Type AC cable shall be supported and secured by which of the following:
    a. Staples and straps
    b. Cable ties
    c. Any of the above
    d. None of the above

23. An installation featuring transverse metallic raceways with electric conductors used to provide access to predetermined cells of a precast cellular concrete floor in order to allow installation of electric conductors from a distribution center to the floor cells is called which of the following:
    a. Cellular concrete floor raceway connector
    b. A raceway cell
    c. The header
    d. A raceway insert

24. Type MI cable is permitted for which of the following uses:
- **a.** In dry locations
- **b.** In wet locations
- **c.** In continuously moist locations
- **d.** All of the above

25. Flat cable assemblies may consist of which of the following:
- **a.** Two or three conductors
- **b.** Four conductors
- **c.** Five conductors
- **d.** All of the above

26. An insulated conductor assembly with fittings and conductor terminations that is completely enclosed in a ventilated protective metal housing is called which of the following:
- **a.** A raceway
- **b.** A busbar
- **c.** A cablebus
- **d.** A busway

27. A cablebus shall not be used for which of the following:
- **a.** Branch circuits
- **b.** Feeders
- **c.** Service cables
- **d.** None of the above

28. Unless identified otherwise, a busway shall not be installed in which of the following locations:
- **a.** Outside
- **b.** In wet locations
- **c.** In damp locations
- **d.** All of the above

29. Branches from busways shall be permitted to use which of the following wiring methods:
- **a.** Type MC and Type AC cable
- **b.** Type RMC and Type FMC conduits
- **c.** Rigid nonmetallic conduit and electrical nonmetallic tubing
- **d.** All of the above

30. Which of the following means shall be installed to remove condensed moisture for low points in a busway run:
- **a.** Drainage plugs
- **b.** Filter drains
- **c.** Either of the above
- **d.** Neither-busway runs are not permitted to have low points

Wiring Methods and Materials

31. If a neutral bus is required, it shall be sized to carry all neutral load current and shall meet which of the following stipulations:

    a. Be of adequate size to carry harmonic currents

    b. Have an adequate momentary rating consistent with any system requirements

    c. Have a short-circuit rating consistent with system requirements

    d. All of the above

32. A sheet metal auxiliary gutter may be used for which of the following applications:

    a. To supplement wiring spaces at switchboards

    b. To enclose overcurrent devices

    c. In outdoor installations only

    d. None of the above

33. Bare copper bars in a sheet metal auxiliary gutter shall not carry current in excess of which of the following:

    a. 1.75 amperes/mm^2 of the cross section of the conductor

    b. 100 amperes/inches^2 of the cross section of the conductor

    c. 1000 amperes/inches^2 of the cross section of the conductor

    d. 600 volts

34. The entire length of a run of sheet metal auxiliary gutter must be supported at which of the following intervals:

    a. Intervals not exceeding 5 feet

    b. Every 10 feet

    c. Intervals not to exceed every 30 feet

    d. None of the above

35. Nonmetallic auxiliary gutters installed outdoors shall be marked indicating which of the following:

    a. "Suitable for exposure to sunlight"

    b. "Suitable for use in wet locations"

    c. Conductor temperature ratings

    d. All of the above

36. The minimum clearance permitted between bare current-carrying metal parts of different potentials that are mounted on the same surface in an auxiliary gutter shall be which of the following:
   a. Not less than 2 inches for parts that are held in the air
   b. Not less than 2 inches
   c. Both of the above
   d. None of the above

37. FMT is prohibited from being used for branch circuits in which of the following applications:
   a. In lengths over 6 feet
   b. For voltage systems of 1000 volts maximum
   c. Where concealed
   d. All of the above

38. The maximum permitted size for flexible metallic tubing is which of the following:
   a. Trade size 1/2 inch
   b. Trade size 3/4 inch
   c. Trade size 1 inch
   d. Trade size 1 1/4 inch

39. ENT must be securely fastened in which of the following manners:
   a. Intervals not exceeding 6 feet
   b. Within 2 feet of each outlet device or junction box where it terminates
   c. In a length not exceeding 6 feet in distance from a fixture terminal connection for tap connections
   d. None of the above

40. The total number of conductors permitted inside 3/8 inch FMT is which of the following:
   a. Four #16 insulated THHN conductors plus one covered equipment grounding conductor of the same size
   b. Three #16 insulated THHN conductors
   c. Six #16 THHN plus one bare equipment grounding conductor of the same size
   d. Eight #16 insulated THHN conductors

41. LFNC shall be permitted for which of the following purposes:
   a. Where flexibility is required for installations
   b. Where protection is required for conductors within the conduit against vapors, liquids, or solids
   c. Both of the above
   d. None of the above

Wiring Methods and Materials

42. NUCC shall be permitted for use in which of the following installation methods:
    a. In cider fill
    b. Attached to concrete walls of any building
    c. Direct buried inside of EMT
    d. None of the above

43. HDPE conduit must be composed of high density polyethylene, which provides which of the following characteristics:
    a. It is resistant to moisture
    b. It is resistant to corrosive agents
    c. Both of the above
    d. None of the above

44. Barring any exceptions or additional listing allowances, RNC shall be securely fastened compliant with which of the following:
    a. Within 3 feet of an outlet box
    b. Within 6 feet of a junction box
    c. Within 2 feet of a device box
    d. All of the above

45. If six-inch RNC is installed, it shall be supported at which of the following spacing intervals:
    a. No farther apart than every 3 feet
    b. No farther apart than every 5 feet
    c. No farther apart than every 6 feet
    d. No farther apart than every 8 feet

46. The approved trade sizes for RMC are which of the following:
    a. Minimum 1/2 inch and maximum size 4
    b. Minimum 1/4 inch and maximum size 3
    c. Minimum 1/2 inch and maximum size 6
    d. Minimum 3/4 inch and maximum size 4

47. Intermediate metal conduit shall be permitted in which of the following installations:
    a. Any type of occupancy
    b. Corrosive environments
    c. Wet locations
    d. All of the above

48. IMC shall be permitted for use as which of the following:
   a. An equipment grounding conductor
   b. An unsupported vertical riser from fixed equipment
   c. Both of the above
   d. None of the above

49. Underground feeder cable shall be prohibited for which of the following installations:
   a. In hoistways
   b. In theaters
   c. In commercial garages
   d. All of the above

50. Underground feeder cable shall be permitted for which of the following applications:
   a. In hoistways
   b. In theaters
   c. In commercial garages
   d. All of the above

51. The minimum bending radius for trade size 4 NUCC is which of the following:
   a. 60 inches
   b. 48 inches
   c. 36 inches
   d. 24 inches

52. Type SE cable shall be permitted in which of the following wiring systems:
   a. Where all of the conductors of the cable are rubber covered
   b. Where all of the conductors of the cable are thermoplastic type
   c. Either of the above
   d. None of the above

53. A nonmetallic extension shall consist of which of the following installation requirements:
   a. A continuous, unbroken length of assembly must be used
   b. The assembly shall not contain splices
   c. The extension assembly can not have exposed conductors between fittings
   d. All of the above

54. Power and control tray cable, type TC, shall be permitted for which of the following applications:
   a. For Class 1 circuits as allowed in Parts II and III of Article 300
   b. For non-power limited fire alarm circuits with compliant copper conductors
   c. For power, lighting, signal, and control circuits
   d. All of the above

Wiring Methods and Materials

55. Unless identified otherwise, TC cables shall not be installed or used as which of the following:
   a. In outdoor locations supported by messenger wire
   b. Outside of a raceway
   c. With insulated conductors size #18 AWG copper
   d. All of the above

56. Unless otherwise prohibited, type NMC cable is permitted for which of the following applications:
   a. In outside masonry block walls
   b. In exposed work
   c. In moist, damp or corrosive locations
   d. All of the above

57. Type NMS cable is permitted for which of the following applications:
   a. Corrosive locations exposed to fumes or vapors
   b. Embedded in masonry
   c. Exposed work in normally dry locations
   d. All of the above

58. Type MI cable shall not be used for which of the following applications:
   a. Where embedded in plaster or concrete
   b. As service cabling
   c. Attached to cable trays
   d. None of the above

59. The maximum allowable distance between supports for RMC is which of the following:
   a. Trade size 1 = 12 feet
   b. Trade size 6 = 20 feet
   c. Trade size 2 = 16 feet
   d. All of the above

60. Metal clad cabling shall be permitted to be used for which of the following applications:
   I. Indoors or outdoors
   II. For services or feeders
   III. For branch circuits
   IV. Direct buried in the earth
   a. All of the above
   b. I, III & IV
   c. I, II, & III
   d. II, III & IV

61. Type IGS cabling which is 1750 kcmil shall not exceed which of the following:
- **a.** 400 amperes
- **b.** 344 amperes
- **c.** 200 amperes
- **d.** 110 amperes

62. The actual outside diameter dimension for IGS trade size 3 cable is which of the following:
- **a.** 73.30 inches
- **b.** 2.886 inches
- **c.** 3.500 inches
- **d.** 89 inches

63. Type FCC cabling shall be approved for which of the following uses:
- **a.** Appliance branch circuits
- **b.** General purpose branch circuits that do not exceed 30 amperes
- **c.** Individual branch circuits that do not exceed 20 amperes
- **d.** All of the above

64. A busway must have a permanent nameplate which indicates which of the following combinations of information:
  I. Rated voltage and rated impulse withstand voltage
  II. Rated continuous current and rated momentary current
  III. Rated frequency and rated 60-Hz withstand voltage (dry)
- **a.** I & II
- **b.** I & III
- **c.** II only
- **d.** All of the above

65. The maximum sized carpet square permitted to cover type FCC cable, connectors, and insulating ends is which of the following:
- **a.** 36 inches
- **b.** 1 foot square
- **c.** 24 inches
- **d.** None of the above

66. Which of the following statements is true for FCC cable systems:
- **a.** Unused cable runs and associated cable connectors may be left in place and energized.
- **b.** New cable connectors must be used to make alterations at new connection points.
- **c.** All of the above
- **d.** None of the above

67. FC cable assemblies are permitted to be used in which of the following applications:
- **a.** As branch circuits to supply tap devices for small appliances not to exceed 30 amps
- **b.** In wet locations
- **c.** In damp locations
- **d.** All of the above

Wiring Methods and Materials

68. Armored cable is permitted for use in which of the following installations:
   a. Dry, damp, and wet locations
   b. Embedded in plaster finish on brick in dry locations
   c. Both of the above
   d. None of the above

69. Nonmetallic boxes shall only be used for which of the following installations:
   a. Concealed knob-and-tube wiring systems
   b. Single-dwelling or two-family wiring systems
   c. Both of the above
   d. None of the above

70. The free space required around each #12 AWG conductor in a box shall be which of the following:
   a. 2.25 inches   **b.** 2 inches
   c. 1.75 inches   **d.** 1.50 inches

71. The maximum number of conductors permitted in a trade size 4 × 1 1/2 inches square metal box includes which of the following:
   a. Seventeen #18 AWG and five #6 AWG
   b. Nineteen #18 AWG and eleven #10 AWG
   c. Twenty-eight #18 AWG and eight #6 AWG
   d. None of the above

72. The maximum number of #14 AWG conductors permitted in a FS single gang 1 1/4 inch box is which of the following:
   a. 9   **b.** 7
   c. 6   **d.** 5

73. Conductors run in parallel in each phase, polarity, neutral, or grounded circuit conductor must comply with which of the following:
   a. Be terminated in the same manner and have the same insulation type
   b. Have the same size circular mil area
   c. Be of the same length and conductor material
   d. All of the above

74. In a 1000 volt system, the distance between where a cable or conductor enters a box and where it exits shall be not less than which of the following:

   a. 10 times the outside diameter, over sheath, of the cable or conductor

   b. 18 times the outside diameter of the cable or conductor

   c. 36 times the outside diameter, over sheath, of the cable or conductor

   d. None of the above

75. In a situation where more than one calculated ampacity could apply to a circuit length, which of the following values shall be used:

   a. The lowest

   b. 25% of the combined ampacity

   c. An average of the combined ampacities

   d. None of the above

76. Raceways that are exposed to a variety of temperatures shall be which of the following:

   a. Grounded
   b. Weatherproof
   c. Bonded
   d. Sealed

77. An assembly of units and associated fittings that form a rigid structural system used to support cables and raceways is considered which of the following:

   a. A wireway
   b. A wiring assembly
   c. A cable tray
   d. A bus

78. A pliable raceway has which of the following characteristics:

   a. It requires a manual or automatic bender to ensure safe bends

   b. It can be bent by hand without requiring any other assistance

   c. It is flexible

   d. None of the above

79. If a nonmetallic surface raceway is used, the installation must meet which of the following conditions:

   a. The building must be used for offices

   b. The building cannot exceed four floors

   c. Both of the above

   d. None of the above

Wiring Methods and Materials

80. The maximum number of current-carrying conductors that shall be used at any cross-section of a wireway is which of the following:
    a. 30
    b. 20
    c. 10
    d. 5

81. The letter "D" indicates two insulated conductors that are installed in which of the following manners:
    a. Laid in divided or separate wireways
    b. Laid in parallel inside an outer nonmetallic covering
    c. Permitted to be connected with flexible connectors
    d. None of the above

82. If there are 7 current-carrying conductors in a raceway, then the individual ampacity of each conductor must be reduced by which of the following:
    a. 70% because of the number of conductors
    b. 25% because all conductors are current-carrying
    c. 80% of the continuous load
    d. None of the above

83. A conduit that enters a box or gutter shall have insulated bushings if the conduit contains which of the following:
    a. #4 AWG conductors
    b. 10 or more conductors
    c. Conductors that are installed in a damp location
    d. #2 or smaller conductors

84. Strut-type channel raceways shall be made of which of the following materials:
    a. Stainless steel
    b. Aluminum
    c. Either of the above
    d. None of the above

85. A receptacle connection made in the white wire of a multiwire circuit shall be made in accordance with which of the following:
    a. The white wire shall be attached to the receptacle box
    b. With a pigtail to the silver terminal
    c. Grounded to a brass screw
    d. All of the above

86. Support of #0 copper conductors in a vertical raceway shall be at intervals not to exceed which of the following:

   a. Every 6 feet
   b. Every 25 feet
   c. Every 50 feet
   d. Every 100 feet

87. In an area with heavy vehicular traffic, rigid conduit shall be buried with a minimum cover of which of the following:

   a. 2 feet
   b. 18 inches
   c. 1 foot
   d. 6 inches

88. IGS cable insulation shall have a nominal gas pressure of which of the following:

   a. 30 pounds per square inch gage
   b. 20 pounds per square inch gage
   c. 30 pounds per inch gage
   d. 20 pounds per inch gage

89. Type NM 600 volt insulated conductors shall be which of the following sizes:

   a. Size 14 AWG through #2 AWG copper-clad aluminum conductors
   b. Size 14 AWG through #2 AWG copper conductors
   c. Size 12 AWG through #4 AWG aluminum conductors
   d. All of the above

90. Type SE service entrance cables are allowed in interior wiring systems with circuit conductors that are all which of the following:

   a. #0 AWG or less
   b. Thermoplastic and rubber-coated
   c. Run in parallel
   d. All of the above

91. A strut-type metallic channel raceway must be secured in which of the following ways:

   a. Within 3 feet of each outlet box or other channel raceway termination
   b. Within 3 feet of each junction box or other channel raceway termination
   c. Secured to the mounting surface with external retention straps located no more than 10 feet apart
   d. All of the above

Wiring Methods and Materials

92. Installations of type FCC systems require which of the following:
   a. A metal top shield over any floor-mounted FCC cable
   b. A bottom shield underneath the FCC cable
   c. Either of the above
   d. Both of the above

93. Securing nonmetallic sheathed cable is not required when the boxes attached meet which of the following requirements:
   a. The box is no larger than 2 1/4 inches × 4 inches mounted in the wall and the cable is fastened within 6 inches of the box
   b. The box is no larger than 2 1/4 inches × 4 inches mounted in the wall and the cable is fastened within 8 inches of the box
   c. The box is no larger than 2 1/2 inches × 4 inches mounted in the wall and the cable is fastened within 12 inches of the box
   d. None of the above

94. The following wiring method is approved to be installed inside a duct used for vapor removal and ventilation of commercial type equipment:
   a. EMT
   b. Nonmetallic sheathed cable
   c. Rigid steel conduit
   d. None of the above

95. When conductors of different systems are installed in a common raceway or cable the derating factors used shall be which of the following:
   a. Applied only to the number of power and lighting conductors
   b. Be increased by one over the total derating factor
   c. Be at least 50% of the highest rated cable
   d. None of the above

96. When two different ampacities apply to adjacent portions of a circuit then the higher ampacity shall be permitted to be used under which of the following conditions:
   a. Beyond the transition point a distance equal to 10 feet
   b. 10% of the circuit length based on the higher ampacity
   c. A or B, whichever is less
   d. Both A and B

97. Screws used to mount knobs shall be long enough to penetrate the wood in accordance with which of the following:
    a. At a depth which is twice the height of the knob
    b. At thickness at least twice the thickness of the screw
    c. At a depth at least one-half the height of the knob
    d. None of the above

98. When calculating the number of conductors in a box, a conductor running through the box shall be counted as which of the following:
    a. Zero
    b. One
    c. Two
    d. Three

99. When installed in raceways, conductors which are #8 AWG or larger shall meet which of the following requirements:
    a. Not be insulated
    b. Be stranded
    c. Both of the Above
    d. Either of the above

100. When a conductor is installed in conduit exposed to direct sunlight and in close proximity to a rooftop, under certain circumstances it can experience which of the following:
    a. A temperature rise of 30 degrees F above the ambient temperature on which the conductor ampacity is based
    b. 10 degrees C above the conductor temperature rating
    c. A swelling in the conductor insulation
    d. Cable stretch

ANSWERS

QUIZ 1—ARTICLE 300

1. **B** Table [314.6(A)]
2. **A** Section [314.16(B)(5)]
3. **B** Section [314.17]: Where entering boxes, etc.
4. **C** Section [314.17(C)]
5. **B** Section [312.20]: Shall be flush with or extend beyond the finished surface
6. **B** Section [314.28(A)(1)]
7. **D** Section [358.26]
8. **D** Section [358.30(A)]

> ▶ **Test** tip
>
> This rule applies to any kind of box (outlet, device, junction, etc.).

9. **C** Section [362.30(A)]
   (NOTE: ENT is the abbreviation for Electrical Nonmetallic Tubing.)
10. **D** Section [348.20 (B)]
11. **D** Section [348.30(A)]
12. **B** Section [344.20(A)]

> ▶ **Test** tip
>
> RMC is the abbreviation for Rigid Metal Conduit

13. **B** Section [314.20]
14. **A** Section [386.10(1)]

> ▶ **Test** tip
>
> Unless an exam question specifically states that it includes a code exception, do not base your answer on any exceptions. The most correct answer is one which does not take exceptions into account, unless you are instructed to do so.

15. **B** Section [312.6(C)]

> ▶ **Test** tip
>
> Don't be fooled into picking "None of the above" answers just because you are not certain of the correct answer.

16. **D** Table [310.15(B)(6)]

> ▶ **Test** tip
> The key element of this question is that the term "single-family" makes it a residential design with a single-phase 120/240 volt 3-wire system.

17. **A** Section [334.30]
18. **A** Section [314.17(C)], Exception

> ▶ **Test** tip
> The phrases "not required" and "so long as" create an "only if" or exception scenario.

19. **D** Section [300.6(A)(1)]
20. **C** Section [352.26]

QUIZ 2—COMPREHENSIVE TEST-NEC® CHAPTERS 1 THROUGH 3

1. **B** Article 90 grants permission to the AHJ for exceptions outside of a building or which terminate immediately inside a building wall only.
2. **B** Section [230.24(B)(1)]
3. **A** [Table 300.5] (NOTE: The required depth is 24 inches under a parking lot, regardless of whether it is covered by concrete or not.)
4. **A** Section [90.2(B)]
5. **C** Section [100] (NOTE: The 2005 definition allows a device to carry or control electric energy, but not utilize energy.)
6. **A** Sections [240.4(B)], [240.6(A)] and Table [310.16]
7. **C** Article 100
8. **B** (NOTE: If the cable is specifically identified otherwise, or if it complies with Section [300.5] or [300.50], it may be direct buried or encased in concrete.)
9. **D** Section [110.16], Note
10. **A** Section [370.4(C)]
11. **B** Section [386.10(1)]
12. **B** Section [90.2(B)] (NOTE: Electrical engineers are not considered an Authority Having Jurisdiction.)

13. **B** Section [344.40(A)]
14. **C** Section [250.118(7)(b)]

> ▶ **Test** tip
> Do not be distracted by answer options which negate the question. Generally speaking, the answer that proves all of the other options false will be listed as "None of the above."

15. **A** Section [250.119(A)], Exception
16. **C** Section [250.92(B)(2) & (4)]

> ▶ **Test** tip
> An answer option of "Either of the above" is very different from "Both of the above." "Both of the above" ties the two answer options together and requires that both conditions must exist. However, the Section which applies to this question allows that ONE of the conditions exists, but does not require that they both exist.

17. **B** Section [250.106]
18. **D** Section [110.26]
19. **A** Article 100

> ▶ **Test** tip
> Although answer B includes part of the definition of a System Bonding Jumper, answer A is the more comprehensive, best answer.

20. **D** Section [334.30]
21. **C** Section [320.10(1)]
22. **C** Section [250.53(F)]
23. **D** Table [110.31]
24. **A** Section [210.19(A)(1)]
25. **B** Table [310.13]

> ▶ **Test** tip
> This cable is allowed in damp locations. Damp is not the same as wet, so only answer B is correct.

26. **D** Section [210.52(E)]

> **► Test** tip
> In your search for the most current answer, consider these factors: A dwelling unit may have more than two exterior doors, so A is not the best answer. Answer B is only partially correct. Answer C, while it totals the same number of outlets as listed in Answer D, is not as comprehensively correct as Answer D.

27. **B** Article 100
28. **A** Section [225.18]
29. **B** Article 100 (NOTE: The definition assumes there are no permanent cooking provisions.)
30. **C** Section [90.9(C)(4)]

QUIZ 3—TRUE / FALSE

#	Ans	#	Ans	#	Ans	#	Ans
1.	B	26.	A	51.	B	76.	A
2.	B	27.	A	52.	B	77.	B
3.	B	28.	B	53.	A	78.	B
4.	A	29.	B	54.	A	79.	B
5.	B	30.	A	55.	B	80.	B
6.	A	31.	B	56.	A	81.	A
7.	A	32.	A	57.	B	82.	B
8.	A	33.	B	58.	A	83.	A
9.	B	34.	A	59.	A	84.	B
10.	A	35.	B	60.	A	85.	B
11.	A	36.	A	61.	A	86.	A
12.	B	37.	A	62.	B	87.	A
13.	A	38.	A	63.	B	88.	A
14.	B	39.	A	64.	A	89.	B
15.	A	40.	B	65.	A	90.	A
16.	B	41.	A	66.	B	91.	A
17.	B	42.	B	67.	B	92.	A
18.	A	43.	B	68.	A	93.	B
19.	B	44.	A	69.	B	94.	A
20.	B	45.	N	70.	A	95.	B
21.	B	46.	N	71.	B	96.	B
22.	A	47.	A	72.	A	97.	A
23.	B	48.	B	73.	A	98.	B
24.	B	49.	A	74.	B	99.	A
25.	A	50.	B	75.	B	100.	A

QUIZ 4—MULTIPLE CHOICE

1.	C	26.	C	51.	A	76.	D
2.	D	27.	D	52.	C	77.	C
3.	D	28.	D	53.	D	78.	B
4.	C	29.	D	54.	D	79.	A
5.	B	30.	C	55.	B	80.	A
6.	D	31.	D	56.	D	81.	B
7.	D	32.	B	57.	C	82.	A
8.	D	33.	C	58.	D	83.	A
9.	A	34.	A	59.	D	84.	C
10.	A	35.	D	60.	C	85.	B
11.	B	36.	B	61.	B	86.	D
12.	A	37.	A	62.	C	87.	A
13.	D	38.	B	63.	A	88.	B
14.	B	39.	D	64.	D	89.	B
15.	D	40.	A	65.	A	90.	B
16.	B	41.	C	66.	C	91.	D
17.	A	42.	A	67.	A	92.	D
18.	B	43.	C	68.	B	93.	B
19.	C	44.	A	69.	A	94.	D
20.	C	45.	D	70.	A	95.	A
21.	A	46.	C	71.	B	96.	C
22.	C	47.	D	72.	C	97.	C
23.	C	48.	A	73.	D	98.	B
24.	D	49.	D	74.	C	99.	B
25.	D	50.	D	75.	A	100.	A

—NOTES—

Chapter 4
EQUIPMENT FOR GENERAL USE

Chapter four of the NEC® illustrates how to apply the principles introduced in the first three chapters to applications involving general equipment. The kinds of equipment covered in chapter four include (in order):

- Flexible cords and cables—Article 400
- Fixture wires—Article 402
- Switches—Article 404
- Receptacles, including cord connectors, and attachment plugs (caps)—Article 406
- Switchboards and panelboards—Article 408
- Lamps, lighting, luminaries—Article 410
- Appliances—Article 422
- Fixed electric equipment for space heaters—Article 424
- Motors, sizing, overcurrent protection, control circuit conductors, motor controllers, disconnecting methods—Article 430
- Air conditioning and refrigeration equipment—Article 440
- Generators—Article 445
- Transformers—Article 450
- Phase converters—Article 455
- Capacitors and components—Article 460
- Batteries—Article 480
- Equipment over 600 volts nominal—Article 490

Article 400 covers general requirements, applications, and construction specifications for flexible cords and flexible cables. Since flexible cords and cables are not considered a wiring method, the applications in Article 400 cover only the items listed in Table **[400.4]**. Section **[400.5]** and Table **[400.5(A)]** list the allowable ampacities for these cords and cables, and Section **[400.6]** describes the mandatory markings permitted. Article 402 covers permanent wiring requirements for fixtures. A fixture is viewed as any device that is permanently attached to a building and considered part of it. Regardless of what the device may be, no fixture wire can be smaller than 18 AWG. Approved fixture wire types are listed in Table **[402.3]**.

> **! Code update SECTION [404.8(B)]**
> The 2002 NEC® stated that the voltage between adjacent snap switches, receptacles, and other such devices could not exceed 300 volts **unless the devices were separated by a permanently installed barrier** between adjacent devices. The 2005 NEC® eliminated the word **permanently** because barriers used in these types of installations are almost always installed in the field. Now the NEC® only requires that barriers be identified and securely installed between adjacent devices.

The requirements for switches of all kinds, including snap or toggle switches, dimmers, fan switches, knife switches, circuit breakers used as switches, and automatic switches are listed in Article 404. Under

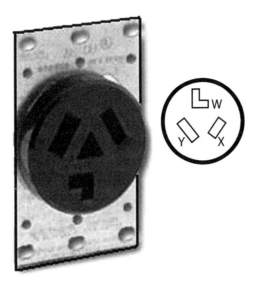

FIGURE 4.1 3 pole 3wire nonground 125/250 v 30 amp receptacle

certain conditions, enclosures for switches and circuit breakers can have splices. The NEC® requires that installations in wet locations, such as areas near showers or pools, include measures to protect against shock, such as using only materials approved for wet locations in order to eliminate seepage. The manners in which switches may be grouped, mounted, marked, grounded, rated, and labeled are outlined. Switching requirements apply to automatic switches used as time clocks and timers, and switches or circuit breakers used as disconnect devices.

Article 406 contains the requirements for receptacles, cord connectors, and attachment plugs (cord caps) grounding and installations. Section **[406.3]** outlines the installation and grounding rules for receptacle outlets, and details for mounting them are provided in Section **[406.4]**. In Section **[406.9(B)(4)]** you will find an example of one of the typical symbols used to identify the termination point for an equipment grounding conductor, which looks like the figure below.

Specific requirements for switchboards, panelboards, and distribution boards that control lights or power circuits are explained in Article 408. These sections of the code are designed to insure installation methods that protect current-carrying conductors from coming in contact with people or maintenance equipment. Section **[408.3(E)]** outlines phase arrangement of these devices. Section **[408.7(III)]** explains that panelboards must be durably marked by the manufacturer with the voltage, current rating, and number of phases at which the equipment has been designed to operate properly. Details of the grounding requirements for panelboards are provided in Section **[408.40].**

FIGURE 4.2 Equipment grounding conductor termination.

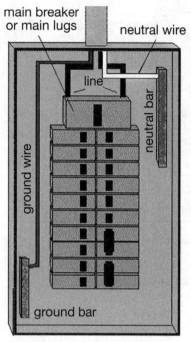

FIGURE 4.3 Electrical panel.

The section of Chapter 4 that covers luminaires, lampholders, pendants, and any of the wiring or equipment that makes up lamping devices begins with Article 410. Specific location installation requirements for luminaires are covered in Section **[410.3(II)]** and a diagram specific to closet installations is provided in figure **[410.8]**. Article 410 lists all types of conditions and installations, including wet locations, and electric appliance requirements are outlined in Article 422. Section **[422.10]** specifies minimum circuit protection and Section **[422.11]** specifies the maximum circuit protection.

> **! Code update SECTION [422.12] (EXCEPTION)**
>
> The 2005 NEC® added an exception to Section [422.12] that allows permanently connected air conditioning equipment to be installed on the same branch circuit as central heating equipment. The previous version of the code dictated that no other equipment could be installed on any dedicated central heating branch circuits.

The requirements for electric heating equipment including heating cable, unit heaters, boilers, central systems, and fixed electric space-heating equipment are listed in Article 424. Section **[424.19]** describes how to provide proper disconnection means for fixed space-heater, heater motor controls, and supplementary overcurrent protection devices. Several means of marking and identifying heating equipment and cables are described in Article 424 IV and Article 424 V. Here you find that each fixed electric heater must have its own nameplate that shows the name and normal operating voltage and watts or amperes rating of

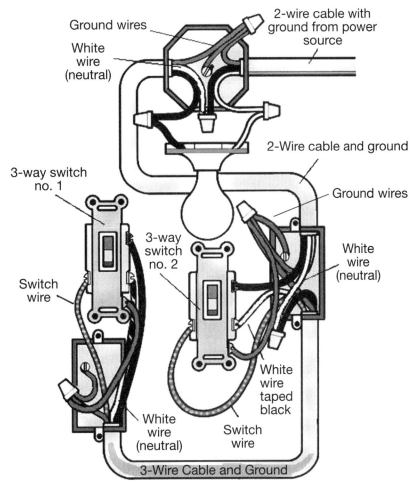

FIGURE 4.4 Luminaire wiring.

FIGURE 4.5 Fused disconnect.

the unit. In addition, Section **[424.35]** lists specific color requirements for the lead wires to fixed electric space heaters; the colors indicate the circuit voltage used and are as follows:

- 120 volts nominal = yellow
- 208 volts nominal = blue
- 240 volts nominal = red
- 277 volts nominal = brown
- 480 volts nominal = orange

Duct heater installations are covered in Article 424 VI; resistance-type boiler provisions are found in Article 424 VII, and electrode-type boiler information is outlined in Article 424 VIII. Article 424 IX explains electric radiant heating panels and panel sets, including how they must be marked and labeled, and the locations in which these types of heaters are prohibited.

> **! Code update SECTION [430.53(C)(6)]**
>
> An addition to [430.53(C)] in the 2005 NEC® is intended to clarify that overcurrent protection for non-motor loads that are on the same equipment or part of a group must comply with the requirements of article 240, part I—part IV. Motor loads and non-motor loads are protected differently, and only the overcurrent protection for the motors is listed in Article 430.

Article 430, which is very extensive, contains the regulations for conductor sizing, overcurrent protection, control circuit conductors, motor controllers, and disconnecting means, while Article 445, which is fairly short and succinct, outlines the requirements for generators. A motor produces rotation when supplied with electricity, so there are many factors and conditions explained in Article 430. Figure **[430.1]** is a quick reference diagram showing where various requirements for each type of equipment may be found in this Article.

> **▶ Test tip**
>
> This chart will save you from needlessly scanning through this Article for information.

There are a number of valuable tables in this section of the code. For example, if you are installing a motor assembly that uses trip coils or relays, instead of fuses, for overload protection, then Table **[430.37]** shows the number and location of overload units required based on the kind of motor and system supply of the unit. It shows that in a two-phase AC motor with a three-wire, two-phase AC supply where one wire is grounded, you have to have two overload units in ungrounded conductors. Table **[430.72]** lists the maximum rating of overcurrent protection devices in amperes.

Suitable motor controllers for all motors are outlined in Article VII, including controller designs, voltage ratings, motor controller enclosure types, which are detailed in Table **[430.91]**. Each motor must have its own individual controller, with the exception of the scenarios listed in Section **[430.87]**. Article 430 XIV is a collection of tables that illustrate such requirements as the full-load current for direct current motors by amperes (Table **[430.247]**) and the full-load currents, in amperes, for single phase alternating current motors (Table **[430.248]**), for two-phase alternating current motors (Table **[430.249]**) and for three-phase alternating current motors (Table **[430.250]**). A conversion table (Table **[430.251(B)]**) delineates the maximum motor locked-rotor current in amperes of various horsepower motors, by phase type.

Types of refrigeration and air conditioning equipment and the installation requirements for each style are found in Article 440. Article 445 covers generator installations. Since a generator produces electricity when rotated, it is essentially the reverse of a motor. Overcurrent protection for two-wire and three-wire generators is explained in Section **[445.12]**. The ampacity of conductors running from generator terminals to the first distribution device that contains overcurrent protection cannot be less than 115 percent of the current rating listed on the generator nameplate, per Section **[445.13]**. The only exception listed is for generators designed to prevent overloading. In this case only, the ampacity of the conductors cannot be less than 100 percent of the current rating on the generator nameplate.

> **! Code update SECTION [445.18]**
> The wording in this section of the 2002 Code was "a disconnect," which indicated a single disconnect. The 2005 wording change allows more than one disconnection means to disconnect a single generator.

Article 450 provides overcurrent protection requirements and the proper connection requirements for transformers and transformer vaults. Part I outlines general requirements, Part II provides specific requirements for different types of transformers, and Part III details transformer vault requirements. The maximum ratings for the overcurrent protection of transformers that are over 600 volts are listed in Table **[450.3(A)]**, and the ratings for transformers that are less than 600 volts are shown in Table **[450.3(B)]**.

> **! Code update SECTION [450.5]**
> A new sentence was added to this section of the 2002 Code. The change specifically addresses Zig-Zag connected transformers (also called T-connected transformers) and states that they cannot be connected to the load side of any system grounding connections. These connections must be made in accordance with [250.24(B), 250.30(A)(1)], and [250.32(B)(2)]. Zig-Zag transformers are a type of autotransformer usually connected to three-phase ungrounded delta system to provide a neutral reference point for grounding, but if you connected this type of transformer to the load side of a three-phase four-wire grounded system, it could interrupt the fault current when a ground fault occurred and damage the autotransformer.

Phase converters, which are the devices used to convert single-phase power to three-phase power and include rotary-phase and static-phase converters, are covered in Article 455. If a phase converter supplies specific fix loads, and the ampacity of the conductor is less than 125 percent of the singe-phase full-load amperes listed on the phase converter nameplate, then the following rule applies:

> Conductor ampacity cannot be less than 125 percent of the sum of the full-load, three-phase current rating of all of the motors and other loads served where the input and output voltages of the phase converter are the same. If the input and output voltages are different, then ampacity cannot be less than the ratio of output divided by the input voltage.

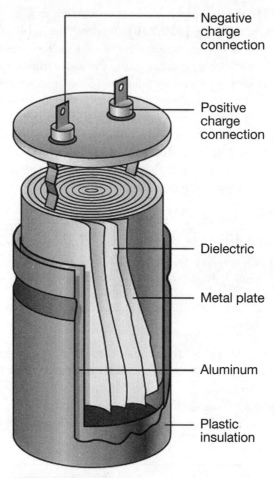

FIGURE 4.6 Capacitor.

Article 460 separates capacitor requirements into two categories: those with 600 volts and less and those over 600 volts. Grounding for capacitor cases was explained in Article 250, but the method for reducing residual voltage in capacitors is found in Section **[460.28]**. Here you are instructed to provide a means of reducing any residual voltage in a capacitor to 50 volts or less within 5 minutes after the capacitor is disconnected from the power source.

Even though resistors and reactors are components of other equipment, they have their own code applications listed in Article 470. The same holds true for stationary storage battery requirements, which are found in Article 480. Finally, the overall requirements for equipment that operates at over 600 volts nominal are laid out in Article 490. This equipment includes circuit breakers, fuses, fuse links and distribution cut-outs, oil-filled cutouts, load interrupter switches, metal-enclosed power switchgear and industrial control assemblies.

> **! Code update SECTION [490.46]**
> This Section was added to Article 490. It requires metal enclosed and metal clad switchgear used as high voltage service equipment and operated at over 600 volts to include a grounding bus that extends into the enclosure or compartment where service conductors are terminated.

With our NEC® chapter four review behind us, it's time again to test your knowledge of equipment installation requirements. This is not a timed test. Like so many of the other chapter review tests in this book, your goal in answering these questions is to identify your strengths and any areas of weakness that you may need to work on.

QUIZ 1—ARTICLE 400

1. Only one disconnecting means shall be allowed to disconnect a single generator.

    a. True   **b.** False

2. For fixed electric space heating equipment consisting of resistance elements with a motor, the branch circuit conductor ampacity and the overcurrent rating of the protective device that supplies the equipment shall not be less than which of the following:

    a. 100% of the total heating equipment load

    b. 125% of the total motor load

    c. 125% of the total load of the motor and the heaters

    d. The combined ampacity of all of the equipment

3. The disconnection means for fixed appliances within a ten-unit apartment building must be in which of the following locations:

    a. Inside each individual dwelling unit

    b. In the main mechanical room of the building

    c. In an accessible common location within the building

    d. Within each individual unit or on the same floor in the building

4. If a GFI breaker is not provided to protect impedance heating, then the secondary voltage cannot exceed which of the following:

    a. 25 volts   **b.** 15 amps

    c. 80 volts   **d.** 600 volts nominal

5. The installation of delta breakers is not allowed in which of the following:

    a. Buss   **b.** Enclosures

    c. Box   **d.** Panelboards

6. When all clearance requirements are met, which type of lighting fixture may be installed in a clothes closet:

    a. A surface mounted fluorescent fixture with an exposed lamp

    b. A surface mounted fluorescent fixture with a completely enclosed lamp

    c. A surface mounted incandescent fixture with a completely enclosed lamp

    d. All of the above

7. A 120 volt, cord and plug type window air conditioner supplied by a general purpose 20 amp branch circuit, cannot have a full-load current that exceeds which of the following:

    a. 10 amps   **b.** 20 amps

    c. 30 amps   **d.** 40 amps

8. A transformer vault that is protected by a halon system shall have a door with a minimum fire rating equal to which of the following:

    a. 1 hour

    b. 3 hours

    c. 6 inches of reinforced concrete

    d. None of the above

9. Circuit conductors that supply power conversion equipment included as part of an adjustable-speed drive system must have an ampacity of which of the following:

    a. At least 125 percent of the motor's full-load current

    b. Not less than 125 percent of the rated input to the power conversion equipment

    c. Not less than 50 percent of the maximum ampacity listed on the equipment nameplate

    d. All of the above

10. The disconnection means for a 400 volt motor circuit must have an ampere rating of which of the following:

    a. At least 125 percent of the motor's full-load current

    b. Not less than 115 percent of the full-load current rating of the motor

    c. 120 amps

    d. Equal to or greater than that the ampacity listed on the motor equipment

11. Lamp tie wires, mounting screws, and clips on light fixture glass must be grounded.

    a. True   **b.** False

12. A 15 amp receptacle that is installed in a wet location must meet which of the following installation requirements:

    a. Be protected from rain or water runoff

    b. Have an attachment plug cap inserted

    c. Have an enclosure that is weatherproof

    d. All of the above

13. Receptacles are considered grounded by which of the following methods:
    a. When the grounding contacts are connected to the equipment grounding conductor of the circuit that supplies the receptacle
    b. If the receptacle ground wire is terminated under a metal screw
    c. When wired to a cord connector
    d. When the grounding contacts have been effectively grounded

14. Luminaires may be used as circuit conductor raceways.
    a. True
    b. False

15. If an electric-discharge lamp has any lamp terminals connected to a circuit over 300 volts, then the terminals shall be considered which of the following:
    a. As having formed a continuous loop circuit
    b. Circuit conductors
    c. Energized
    d. A raceway

16. Fixed outdoor snow melting and de-icing equipment is considered to have a continuous load.
    a. True
    b. False

17. Installations for the electrical heating of a pipeline must:
    a. Be protected from physical damage
    b. Include caution signs posted at frequent intervals along the pipeline
    c. Be identified as being suitable for the chemical, thermal and physical environment of the installation
    d. All of the above

18. If a device is permanently attached to a building and considered part of it, then the minimum approved wire size shall be which of the following:
    a. 12 AWG
    b. 16 AWG
    c. 14 AWG
    d. 18 AWG

19. The allowable ampacity of an 18 AWG fixture wire is which of the following:
    a. 18
    b. 15
    c. 8
    d. 6

Equipment for General Use   113

20. An industrial control panel supply conductor shall have an ampacity of which of the following:

  a. No less than 125% of the full-load current rating of all resistance heating loads and no more than 125% of all combined continuous loads

  b. No less than 125% of the full-load current rating of all resistance heating loads plus 125% of the full-load current rating of all other connected motors based on their duty cycle if they are all in operation at the same time

  c. No less than 125% of the full-load of two or more components of a systematic assembly

  d. Not to exceed the ampacity listed for all resistance heating equipment and connected motor nameplates

21. For a 2-phase AC motor connected to a three-wire, two-phase ungrounded conductor supply system, where fuses are not used for overload protection, overload protection relays must be installed in which of the following manners:

  a. One relay on either ungrounded conductor, for a total of one

  b. One relay on each ungrounded conductor, for a total of two

  c. One relay in each phase, for a total of two

  d. One relay per wire, for a total of three

22. The full-load current for a single-phase alternating current 115 volt, 1 horsepower motor is which of the following:

  a. 8 amps   **b.** 15 amps

  c. 16 amps   **d.** 20 amps

23. In order to guard exposed live motor parts and controllers operating at 50 volts or higher from accidental contact, which of the following methods must be used:

  a. Equipment is installed in a room that is only accessible by a qualified person

  b. Equipment is installed on a balcony that is elevated enough to prohibit access by unqualified people

  c. Equipment is installed or mounted at least 8 feet above floor level

  d. All of the above

24. A generator must be equipped with a disconnection means that will entirely disconnect the generator and all protective devices and control apparatus from the circuits supplying the generator.

  a. True   **b.** False

25. A when a transformer is used to create a three-phase, four-wire distribution system from a three-phase, three-wire ungrounded system, the transformer must meet which of the following installation requirements:

   a. The transformer must not be switched

   b. The transformer must be directly connected to the ungrounded phase conductors

   c. The transformer shall not be provided with overcurrent protection that is independent from the main switch and common-trip overcurrent protection for a three-phase, four-wire system

   d. All of the above

26. A motor controller is required to open all conductors on a motor.

   a. True   b. False

27. Barring any exceptions or special conditions, the enclosure for a motor controller shall be permitted to be used as a junction box.

   a. True   b. False

28. The maximum rating for heating panels used for de-icing is 120 watts/square foot.

   a. True   b. False

29. Excess leads on heating cables must be secured and may be cut as needed.

   a. True   b. False

30. Heating cables installed on ceiling boards shall be secured every 16 inches.

   a. True   b. False

31. On a 120 volt space heater, the non-heating cable leads are yellow.

   a. True   b. False

32. Conductors installed above a heating ceiling are considered to be in a 40(degrees C ambient temperature space.

   a. True   b. False

33. As long as other disconnection means are provided, then unit switches may disconnect a heater.

   a. True   b. False

34. For a wall-mounted electric oven, a cord-and-plug is considered the disconnection means.

   a. True   b. False

Equipment for General Use

35. A type SP-1 flexible cord shall be used to connect an electric dishwasher.

    a. True   **b.** False

36. An electric toaster may have live parts that are exposed to contact.

    a. True   **b.** False

37. Flexible cords are permitted only to supply portable appliances.

    a. True   **b.** False

38. In a lead acid battery, the nominal voltage is 1.75 volts per cell.

    a. True   **b.** False

39. Protection of the live parts of a generator that operates at 30 volts is required if the generator is accessible to unqualified persons.

    a. True   **b.** False

40. Within 30 seconds of being disconnected from the power supply, a capacitor which operates at 300 volts must be discharged to 50 volts.

    a. True   **b.** False

41. A 40 amp branch circuit may supply up to five electric space heaters installed in a two-family dwelling.

    a. True   **b.** False

42. Under no circumstances shall a hermetic, motor compressor have an overcurrent protection device rated more than 225% of the rated load current.

    a. True   **b.** False

43. All appliances must have a disconnection means.

    a. True   **b.** False

44. A fixed resistive-type space heater shall have branch circuit conductors that are rated at 110% of the heat load.

    a. True   **b.** False

45. Barring any exceptions, phase A must have the highest voltage to ground in a high-leg delta arrangement on a switchboard with bus bars.

    a. True   **b.** False

46. A 200 amp panelboard shall be permitted to supply a 200 amp non-continuous load.
    a. True   b. False

47. In a lighting and appliance branch circuit panelboard, at least 10% of the overcurrent devices shall be rated at 30 amps or less.
    a. True   b. False

48. A transformer rated over 35,000 volts shall be installed in a vault.
    a. True   b. False

49. In a panelboard that is not used as service equipment, the grounding terminal bar is not connected to the neutral bar.
    a. True   b. False

50. A knife switch rated more than 1200 amps at 250 volts shall be used only as an isolating switch.
    a. True   b. False

ANSWERS

QUIZ 1—ARTICLE 400

1. **B** Section [445.18] (NOTE: The 2005 edition allows for more than one disconnection means.)
2. **C** Section [424.3(B)]
3. **D** Section [422.34(A)]

> ► **Test** tip
> Keeping in mind that you are looking for the most correct answer, you can see that Answer A is part of Answer D, and although being located on the same floor may mean the disconnection means is in a common, accessible location, Answer C is not as comprehensively correct as Answer D.

4. **C** Section [426.32]
5. **D** Section [408.36(E)]
6. **D** Section [410.8(C)]
7. **A** Section [440.62(C)] (NOTE: This section requires that that this type of installation not exceed 50% of the rating for the circuit. 20 amp × 0.5 = 10 amps.)
8. **A** Section [450.43(A)], Exception
9. **B** Section [430.122(A)]
10. **B** Section [430.110(A)]
11. **B** Section [410.18(A)]

> ► **Test** tip
> Some answers are just common sense. Have you ever grounded the clips that hold the glass onto a light fixture?

12. **C** Section [406.8(B)(1)]

> ► **Test** tip
> Answer D might seem correct at first glance; however, the requirement for a receptacle of this amperage in a wet location states it must have a weatherproof enclosure "whether or not the attachment plug cap is inserted."

13. **A** Section [406.3(C)]

14. B Section [410.31] prohibits this unless the fixture is listed and marked for use as a raceway. This condition was not included in the question.
15. C Section [410.73(C)]
16. A Section [426.4]
17. D Sections [427.10, 11, 12 and 13]
18. D Section [402.6]
19. D Table [402.5]
20. B Section [409.20]
21. C Table [430.37]
22. C Table [430.248]
23. D Section [430.232(1), (2) and(3)]
24. A Section [445.18]
25. D Section [450.5(A) and (A1)]
26. B Section [430.84]
27. B Section [430.10(A)]
28. A Section [426.20(B)]
29. B Section [424.43(C)]
30. A Section [424.41(F)]
31. A Section [424.35]
32. B Section [424.36]
33. A Section [424.19(C)]
34. B Section [422.16(B)(3)], and [422.33(A)]
35. B Section [422.16(B)] Table [400.4]
36. A Section [422.4]
37. B Section [400.7(A)]
38. B Section [480.2]
39. B Section [445.14]
40. B Section [460.8(A)]
41. B Section [424.3(A)]
42. A Section [440.22(A)]
43. A Section [422.30]
44. B Section [424.3(B)]
45. B Section [408.3(E)]
46. A Section [408.36(A)]
47. B Section [408.34] and [408.34(A)]
48. A Section [450.21(C)]
49. A Section [408.40]
50. A Section [404.13(A)]

Chapter 5
SPECIAL OCCUPANCIES AND CLASSIFICATIONS

Hazardous locations and conditions and special situations and structures are explained in Chapter five of the NEC. Environments that pose fire or combustion hazards are listed in Articles 500–510. Requirements covering specific types of facilities that pose additional hazards, such as bulk storage plants or motor fuel dispensing locations, are explained in Articles 511–516. Codes regarding facilities with evacuation issues, such as movie theaters, hospitals, or even carnivals, are explained in Articles 517–525. The hazards involved with specific building types, including mobile homes and floating structures, and the electrical requirements for these buildings are found in Articles 545–553. Marina and boatyard regulations are listed in Article 555, and temporary installations are covered in Article 580.

Article 500 provides a general description of hazardous locations that are classified by the NEC. Sections **[500.5]** and **[500.6]** define NEC® Class and Division requirements for electrical equipment used in locations where explosion hazards or risk of fire may exist because of factors such as flammable gases or vapors, liquids, combustible dust, or ignitable materials.

> **! Code update SECTION [500.5(C)(2)(1)]**
>
> This change makes clearer the description of Class II, Division 2 locations with three different examples. Class II, Division 2 locations are classified as follows:
>
> 1. Locations where some combustible dust is normally in the air but where abnormal operations may increase the suspended dust to ignitable or explosive levels.
> 2. Locations where combustible dust accumulations are normally not concentrated enough to interfere with the operation of electrical equipment unless an "infrequent equipment malfunction" occurs that increases the level of dust suspended in the air.
> 3. Locations where combustible dust concentrations in or on electrical equipment may be sufficient to limit heat dissipation or that could be ignited by failure or abnormal operation of electrical equipment.

A variety of airborne environmental conditions that require classification are listed in Article 500. Class I covers locations specified in Sections **[500.5(B)(1)]** and **[500.5(B)(2)]**—locations where flammable gases or vapors are, or could exist in the air in high enough quantities that they could produce explosive or ignitable mixtures. Section **[500.5(B)(1) FPN 1]** provides examples of locations usually included in Class I, such as the following:

1. Where volatile flammable liquids or liquefied flammable gases are transferred from one container to another
2. Interiors of spray booths and areas in the vicinity of spraying and painting operations where volatile flammable solvents are used
3. Locations containing open tanks or vats of volatile flammable liquids
4. Drying rooms or compartments for the evaporation of flammable solvents
5. Locations with fat and oil extraction equipment that uses volatile flammable solvents
6. Portions of cleaning and dyeing plants where flammable liquids are used
7. Gas generator rooms and other portions of gas manufacturing plants where flammable gas may escape

8. Pump rooms for flammable gas or for volatile flammable liquids that are not adequately ventilated

9. The interiors of refrigerators and freezers where flammable materials are stored in open or easily ruptured containers

10. All other locations where ignitable concentrations of flammable vapors or gases are likely to occur in the course of normal operations

Class I lists groups of air mixtures that are not oxygen enriched. These include the following:

- Group A—Acetylene
- Group B—Flammable gas, flammable vapors that produce liquid, and combustible vapor mixed with air that could burn or explode with a maximum experimental safe gap (MESG) value equal to or less than 0.45 mm, or with a minimum igniting current ratio (MIC ratio) equal to or less than 0.04.
- Group C—Flammable gas, flammable vapors which produce liquid, and combustible vapor mixed with air that could burn or explode with a MESG value greater than 0.45 mm but less than 0.75 mm, or a MIC ratio greater than 0.04 but less than 0.80.
- Group D—Flammable gas, flammable vapors that produce liquid, and combustible vapor mixed with air that could burn or explode with a MESG value greater than 0.75, or a MIC ratio greater than 0.80.

Class II locations continue the group classifications for environments affected by combustible dusts, as follows:

- Group E—atmospheres containing combustible metal dusts whose particle abrasiveness, size, or conductivity poses hazards in the use of electrical equipment.

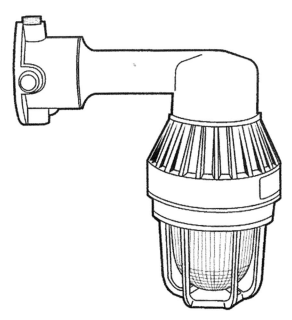

FIGURE 5.1 Hazardous location luminaire.

- Group F—atmospheres containing combustible carbonaceous dusts, such as coal and charcoal, which present an explosion hazard.

- Group G—atmospheres containing combustible dusts that are not already included in Group E or F, such as wood or plastic dust, grain or chemical dust.

The approved protection techniques for these group locations are found in Section **[500.7]**.

Article 501 states that a Class I hazardous (classified) location is an area where flammable gases or vapors may be present in quantities sufficient to produce an explosive or ignitable mixture. Article 501 contains installation requirements for these locations, including wiring methods, equipment connections, and use of devices. Equipment installations included under this classification are:

- General power distribution

- Transformers and capacitors—Section **[501.2]**

- Meters, instruments and relays—Section **[501.3]**

- Wiring methods—Section **[501.4]**

- Seals and drainage requirements—Section **[501.5]**

- Controls, such as covers, switches, fuses, circuit breakers, or motor controllers—Section **[501.6]**

- Transformers and Resistors—Section **[501.7]**

- Motors and generators, load requirements, luminaires, flexible cords, receptacles and conductor insulation—Sections **[501.8–501.13]**

- Alarms, signaling, alarm, remote control and communications systems—Section **[501.14]**

- Surge protection in these hazardous areas—Section **[501.17]**

Article 502 specifically addresses conditions and locations that contain combustible dust in the air, rather than just the presence of combustible materials. A Class II, Division 1 location is a one in which combustible dust is present in the air under normal operating conditions in high enough quantities that it could produce explosive or ignitable mixtures. Additionally, if a mechanical failure or abnormal operation of machinery or equipment might produce combustible dust, then those locations or facilities would fall under Class II, Division 1.

A Class II, Division 2 location is defined as one where combustible dust is present in large enough quantities to produce explosive or ignitable mixtures due to the abnormal operation of systems. Class II, Division 2 also covers situations in which combustible dust accumulations are present but are normally insufficient to interfere with the normal operation of electrical equipment, unless an infrequent equipment malfunction occurs. Finally, combustible dust accumulations on, in, or in the vicinity of the electrical equipment that could interfere with the safe dissipation of heat from electrical equipment, or which could be ignited by an electrical equipment failure are also included in this section.

Article 503 lists Class III locations as areas where fibers or flyings could be easily ignited. The classification standard that separates these locations from the previous ones is that if you don't have enough fibers or flyings in the air to produce an ignitable mixture but they *are present*, then it is a Class III area. Typical examples of a Class III location would be a textile or saw mill or a clothing manufacturing plant. Class III, Division 2 locations are a sub classification where ignitable fibers are stored or handled rather than created in the process of manufacturing.

Article 504 discusses electrical apparatus with circuits that are not necessarily intrinsically—or fundamentally—safe in and of themselves, but which affect the energy and power flow in safely designed circuits. Two examples of these types of systems are provided in Section **[504.2(1) & (2)]**. Section **[504.2]** defines a "simple apparatus" as an electrical component of simple construction that doesn't generate more than 1.5

Special Occupancies and Classifications

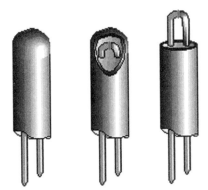

FIGURE 5.2 Thermocouples.

volts, such as thermocouples, photocells, switches, junction boxes, or LED's. Section **[504.10(B)(2)]** provides a calculation method for determining if a simple apparatus is within the maximum allowable surface temperature. Wiring, grounding and bonding requirements for intrinsically safe apparatus are also found in Article 504.

Now, if the classes and grouping of hazardous locations were just way too confusing for you, Article 505 provides zone classifications as an alternative to the division classifications structure listed in Article 500. This article defines such points as combustible gas detection systems (Section **[505.2]**), encapsulation, flame proofing, increased safety, intrinsic safety, purged and pressurized equipment, and types of protection. Types of protection designations are listed in Table **[505.9(C)(2)(4)]**.

Article 511 addresses commercial garages where vehicles that burn volatile liquids, such as gasoline or propane, are stored or repaired. This article lists seven types of unclassified locations and five different kinds of classified areas. Special equipment, such as battery charging equipment, and electric vehicle charging equipment are listed in Section **[511.10]**. Section **[511.12]** requires that all 125-volt, single phase 15 amp and 20 amp receptacles installed in areas where electric diagnostic equipment, electric hand tools, and portable lighting equipment will be used and that ground-fault circuit interrupter protection be installed.

> **! Code update SECTION [511.4(A)(1)]**
>
> In the 2002 NEC, raceways embedded within the earth (soil) or concrete underneath Class I locations were considered to be in a Class I location. This classification and the exception permitting nonmetallic rigid conduit with 24 inches of RMC or IMC risers was deleted in the 2005 edition because of a precise examination of the definition of a Class I area as one with a mixture of flammable gases or vapors in the air. Raceways embedded in the ground or in masonry, aren't exposed to enough air to support combustion.

Article 513, which deals with aircraft hangars, is similar to 511 in concept and approach because aircraft burn volatile liquids also. This article lists mandatory warning sign wording, "KEEP 5 FT. CLEAR OF AIRCRAFT ENGINES AND FUEL TANK AREA," which is required for certain aircraft hangar locations, such as at mobile stanchions (Section **[513.7(F)]**) and mobile battery chargers (Section **[513.10(B)]**) and mobile servicing equipment (Section **[513.10 (D)]**).

Facilities where fuel is dispensed from storage tanks into the fuel tanks of vehicles are covered in Article 514. Here you will find Table **[514.3(B)(1)]**, which illustrates how to classify a motor fuel dispensing area based on the equipment it contains. Article 515 covers rules applying to bulk storage plants, and like the previous chapter, it also provides a chart, Table **[515.3]**, that lists how to classify bulk storage locations based on the extent to which the equipment in them is exposed to various hazardous conditions.

> **! Code** update SECTION **[517.14]**
> The 2005 change to this section alters the wording to "two or more." Additionally, continuous copper conductors are permitted to be "broken" where required to connect them to the equipment grounding buses. These changes are intended to make sure that grounded components of all equipment in an individual patient's vicinity have the same ground potential regardless of the source of the circuit supplying the equipment. Furthermore, they must be bonded together by a continuous copper conductor that is insulated #10 AWG or larger. The same bonding technique is also required for panelboards supplied from separate transfer switches on the emergency system if the panelboards serve the same patient vicinity.

Health care facilities have very specific types of equipment, so a portion of the code is written specifically for these facilities. Article 517 is primarily focused on those parts of health care facilities where patients are examined and treated, whether the facility is permanent or mobile, such as a mobile MRI unit. This article does not provide the wiring and installation requirements for any business office, waiting rooms, or doctor or patient sleeping rooms. Article 512 is broken down into seven parts. The first part includes definitions such as electrical life-support equipment, exposed conductive surfaces and a critical branch which is a subsystem of the emergency system. Parts II and III list requirements unique to hospitals and include requirements on how to:

- Minimize electrical hazards by keeping the voltage potential between patients' bodies and medical equipment low—Section **[517.11]**
- Maximize the physical and electromagnetic protection of wiring by requiring a metal raceway—Section **[517.13(A)]**
- Minimize power interruptions—Section **[517.17]**

Section **[517.14]** requires that equipment grounding terminal buses of essential branch-circuit panelboards that serve the same individual patient vicinity have to be bonded together with a continuous, insulated conductor that cannot be any smaller than 10 AWG. Section **[517.17(B)]** requires an additional step of ground fault protection in the next level feeder disconnects that are downstream, towards the load. It then goes on to list, in Section **[517.17(B)(1)–(3)]** what and where these additional levels *may not* be. In general areas, such as patient bed locations, each bed must by supplied by at least two branch circuits, one from the emergency system and one from the normal power system per the requirements in Section **[517.18]**, and must have at least four receptacles per Section **[517.18(B)]**. Patient beds located in critical care areas listed in Section **[517.19]** must still have at least two branch circuits, with one or more from the emergency system and one or more from the normal power system; at least one of the branch circuits on the emergency system has to supply an outlet at each individual bed location. Each patient bed in a critical care unit must have a minimum of six hospital grade receptacles per Section **[517.19(A)]**, and at least one of them must be connected to either the normal system branch circuit or an emergency system branch circuit that is supplied by a different transfer switch than the rest of the receptacles at that bed location per Section **[517.19(B)(2)]**.

There is a general diagram of the minimum requirements for transfer switch arrangements provided in Table **[FPN 517.30 No. 1]**, and a detailed diagram is provided in Table **[FPN 517.30 No. 2]** for the minimum requirements for transfer switch arrangement in essential systems that are 150 kVA or less. Section **[517.33(A)(1)–(9)]** gives a detailed explanation of the areas and functions that have illumination, fixed equipment, special power circuits or selected receptacles supplied by critical branch wiring. These include infant nurseries, nurses' stations, nurse call stations, coronary care units, and intensive care wards. Section **[517.34(A)(1)–(6)]** describes the types of equipment that must be connected to an alternate power source, such as compressed air systems, central suction systems and controls for medical or surgical functions, and smoke control systems. Installations specific to nursing homes and limited care facilities are listed in Section **[517.40]**. They include a general diagram of the minimum requirements for transfer switch arrangement for a nursing home or limited care facility in Table **[FPN 517.41 No. 1]** and a detailed diagram for the minimum requirements for 150 kVA or less transfer switch arrangement for these facilities in Table **[FPN 517.41 No. 2]**.

Part IV lists the requirements for gas anesthesia stations. The first thing this part establishes is the hazardous, classified locations associated with anesthetizing areas, and wiring and equipment requirements. Wiring methods for anesthesiology areas that are not hazardous are provided in Section **[517.61(C),** and grounding requirements begin in Section **[517.62]**. Part V addresses X-ray installations, specifically ampacity requirements and protection for the branch circuits. The ampacity requirements for diagnostic equipment are found in Section **[517.73(A)]**. This section requires that branch circuit conductors and overcurrent protection devices be at least 50 percent of the momentary rating or 100 percent of the long-term rating of the attached equipment, whichever is greater. Section **[517.73(A)(2)]** is very specific about supply feeder ampacity and current ratings for overcurrent protection for two or more X-ray units—they cannot be less than 50 percent of the momentary rating of the largest unit plus 25 percent of the momentary rating of the next largest unit *plus* 10 percent of the momentary ratings of each additional unit. If two or more X-ray units are used simultaneously, then the supply conductor and overcurrent protection devices must be equal to 100 percent of the momentary demand rating of each of the units.

Part VI provides requirements for low-voltage communications systems, such as fire alarms and intercoms. Part VII provides requirements for isolated power systems. Isolated power circuits must be controlled

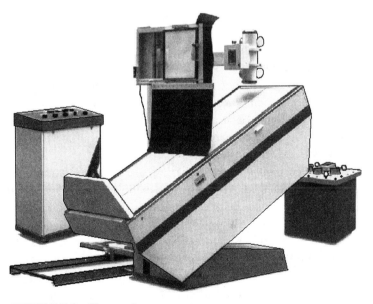

FIGURE 5.2 X-ray unit.

by a switch with a disconnecting pole in each isolated circuit conductor so that the power can be simultaneously disconnected, as described in Part VII, Section **[517.160(A)(1)]**.

Areas classified for assembly are defined in Article 518. If a portion of a building is intended for the assembly of 100 or more people, then certain criteria apply as listed in Sections **[518.1]** and **[518.2(A)]**.

> **! Code** update SECTION [518.2(B)]
> Section [518.2(B)] in the 2005 NEC® was altered to clarify that only those parts of a multioccupancy building that meet the 100-person or more criteria set in Section [518.1] are considered assembly occupancies. Smaller meeting locations are classified as part of other occupancies.

While Article 518 appears to mostly reference requirements in other Articles, it does have several of its own. For example, if you want to install nonmetallic tubing or conduit in spaces with a finish rating, you must satisfy one of two conditions in Section **[518.4(C)(1) & (2)]**, which are as follows:

1. That the tubing or conduit be installed concealed in walls, floors and ceilings where these structures provide a thermal barrier equal to a 15-minute finish rating that complies with the requirements for fire-rated assemblies

2. That tubing or conduit can only be installed above suspended ceilings if those spaces provide a thermal barrier equal to a 15-minute finish rating that complies with the requirements for fire-rated assemblies

Additionally, Section **[518.5]** states that portable switchboards and power distribution equipment can only get power from outlets that are listed to supply adequate voltage and ampere ratings for the devices. These outlets must have overcurrent protection devices and they cannot be accessible by the general public. Also, the neutral for feeders for solid-state, 3-phase, 4-wire dimming systems are considered by this section of the code as current-carrying conductors.

Section **[520.1]** includes a listing of the structures that are considered theaters, audience areas, and performance locations. Article 520 is very detailed and extensive and includes unique installation requirements.

Wiring for carnivals, fairs, circuses, and similar events is covered in Article 525. These locations are similar to places of assembly, except that these locations are temporary and use portable wiring. Additionally, this article includes wiring for amusement rides and attractions that use electricity. Section **[525.5(B)]** requires that these cannot be located any closer than 15 feet in any direction from overhead power lines. If those lines are over 600V, the attractions cannot be located under them. You know that a fairground includes numerous rides, and according to Section **[525.11]**, where there are multiple services or separately derived systems supplying these attractions, then all power sources that are less than 12 feet apart have to be bonded to the same grounding electrode system.

Equipment and wiring installed in agricultural buildings are at risk of damage from two specific factors: dust and corrosion. Dust increases the risks of both fire and explosion. Dust from materials such as hay, grain and fertilizer is highly flammable. Manure from farm animals can create corrosive vapors that can damage mechanical equipment or cause electrical equipment to fail. For these reasons, Article 547 provides requirements for dealing with dust and corrosion. Another factor to consider in agricultural buildings is water, which can be present from not only watering livestock but also from cleaning equipment. Therefore, Article 547 also includes requirements for dealing with wet and damp environments. Section **[547.9]** requires site-isolating devices at the point of distribution where two or more agricultural buildings are supplied.

Special Occupancies and Classifications

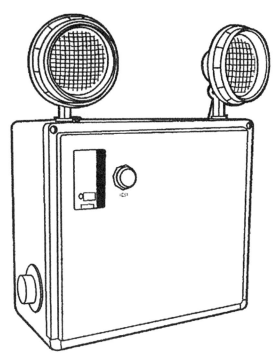

FIGURE 5.4 Emergency unit.

Article 550 recognizes the high incident rate of fires in mobile homes while also providing installation guidelines for mobile or manufactured units, such as construction trailers, that are not used as dwellings. Section **[550.4(A)]** explains that non-dwelling mobile home units are excluded from the 100A minimum service requirement and the number of circuits required in the article for mobile homes used as dwellings. The code in this article recognizes that there is a difference between a mobile home and a manufactured home and explains the distinction in Section **[550.0]**. Because of these differences, some installation requirements vary; for example, you cannot locate service equipment on a mobile home, but you can install service equipment on a manufactured home (provided you meet seven conditions). Grounding methods established in Article 250 are not sufficient for mobile homes, so grounding bus requirements unique to both mobile homes and manufactured homes are explained in Section **[550.16]**.

You might think that an article on recreational vehicles and vehicle parks would only need to include weather-proof receptacles for plug-and-go connection, but Article 551 also includes sections provided for RV manufacturers. Recreational vehicles have voltage converters and low-voltage wiring that can only be copper types, such as GXK, HDT, SGT, SGR and SXL. As an electrician, your requirements really get started in Article 551 Part VI, which requires that you understand how to calculate loads and apply the demand factor, based on the number of RV sites in a recreation vehicle park, so a listing of demand factor percentages is provided in Table **[551.73]**. Receptacle configurations for recreational vehicles are outlined in Section **[551.81]**.

Water levels vary; the ocean has a tide and the depth of lakes and rivers changes depending on elements like rain and evaporation. Article 555 addresses these issues as they relate to electrical installations for marinas and boatyards. When you think water, you have to consider damp and wet conditions as well as reflected sunlight off the water and increased temperatures that can affect insulation factors. Finally, there are additional contributing factors, such as increased abrasion, and the presence of oil, gasoline, diesel fuel, acids and chemicals. Since docking a boat involves a degree of uncontrollable movement, electrical equipment on and

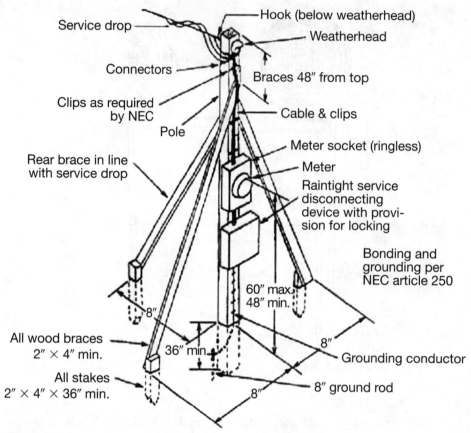

FIGURE 5.5 Temporary installation.

around docks must meet certain spatial requirements so they don't interfere with mooring lines or masts. Article 555 describes the concept of the electrical datum plane, which is basically a "do not enter" zone where no electrical equipment can be installed. Section **[555.19(4)]** addresses the different rating requirements for shore power to boats. Single receptacles cannot be rated less than 30 amperes. Those outlets that range between 30 amperes and 50 amperes must be locking and grounding receptacles, and for 60 or 100 amp receptacles, you have to use pin and sleeve type.

Article 590 is focused on temporary installations. You might think at first that temporary wiring requires lower installation standards than other wiring. However, while the same basic rules of workmanship, ampacity and circuit protection apply to temporary installations, there are a variety of detailed requirements unique to these systems. For example, all receptacles must be of the grounding type. So how is a temporary installation different? In one sense, it does meet a lower standard. For example, you can use NMC rather than the normally required raceway-enclosed wiring without height limitation. And you don't have to put splices in boxes. There are also time limits to temporary installations. For example, if a temporary installation is for holiday displays, it can't last more than 90 days. A time factor comes into play when considering the lifespan of the insulators used in temporary wiring.

With all of these different types of occupancies, hazardous conditions, and various installation requirements behind you, its time once again to test your skills. Since there is so much information in chapter five of the NEC, and because the numerous conditions each have unique requirements, you may need to refer to your code book as you take this quiz.

Special Occupancies and Classifications

QUIZ I

1. Carnival rides cannot be located any closer in any direction from overhead power lines than which of the following:

   a. 12 feet
   b. 15 feet
   c. 18 feet
   d. 20 feet

2. Each patient bed in a hospital critical care unit must have at least how many of the following:

   a. 2 duplex receptacles
   b. 4 single or duplex receptacles
   c. 4 hospital-grade receptacles connected to an emergency system branch circuit
   d. 6 hospital-grade receptacles

3. In places of assembly, the neutral for feeders for solid-state, 3-phase, 4-wire dimming systems are considered to be current-carrying conductors.

   a. True
   b. False

4. Sealed cans of gasoline may be stored in which of the following locations:

   a. Class 1, Division 1
   b. Class 1, Division 2
   c. Class 2, Division 1
   d. Class 2, Division 2

5. A hospital's essential electrical system must be comprised of which of the following two separate systems:

   a. A normal system and a life safety system
   b. An essential system and a non-essential system
   c. A life safety system and an addressable fire alarm system
   d. An emergency system and an equipment system

6. A receptacle installed to provide electric power to a recreational vehicle must comply with which of the following configurations:

   a. 15 amps: 125 volt, 15 amperes, 2-pole, 4-wire grounding type for 120-volt system
   b. 20 amps: 125 volt, 20 amperes, 3-pole, 4-wire grounding type for 120/240 volt system
   c. 30 amps: 125 volt, 30 amperes, 2-pole, 4-wire grounding type for 120-volt system
   d. 50 amps: 125 volt, 50 amperes, 3-pole, 4-wire grounding type for 120/240 volt system

7. A processing room in a perfume manufacturing plant that houses fat and oil extraction equipment is considered which of the following:

   a. Unoccupied space, which requires GFCI outlets

   b. A Class 1 location

   c. A bulk storage location in which continuous copper conductors are permitted to be broken where required to connect it to the equipment grounding buses

   d. None of the above

8. A saw mill is an example of a Class III location.

   a. True
   b. False

9. In an aircraft hangar, there must be a minimum clear space from aircraft engines and fuel tank areas of which of the following:

   a. 5 feet

   b. 6 feet

   c. 10 feet from engines and 6 feet from fuel tanks

   d. 15 feet from both the engines and the fuel tanks

10. Receptacles that provide shore power at marinas must be rated at a minimum of which of the following:

    a. 15 amps
    b. 20 amps
    c. 30 amps
    d. 50 amps

11. Of the hospital-grade receptacles installed in a patient bed location in a critical care unit, at least one of them must be connected to either the normal system branch circuit or an emergency system branch circuit that is supplied by a different transfer switch than the rest of the receptacles at that bed location.

    a. True
    b. False

12. The minimum service size per mobile home site in a mobile home park is which of the following:

    a. 12,000 volts
    b. 15,000 volts
    c. 16,000 volts
    d. 20,000 volts

13. Electrical apparatus with circuits that are not necessarily intrinsically, or fundamentally, safe in and of themselves, but which affect the energy and power flow in safely designed circuits are referred to as which of the following:

    a. Intrinsically safe apparatus
    b. Simple apparatus
    c. Associated apparatus
    d. Alternate apparatus

Special Occupancies and Classifications

14. Electric diagnostic equipment installed in a commercial garage requires which of the following:
    a. A 15 amp or 20 amp receptacle with ground-fault circuit interrupter protection
    b. A 125 volt, single phase 15 amp or 20 amp GFCI receptacle
    c. That the location be considered Class II
    d. A separate ground fault and a 30 amp duplex receptacle

15. 1200 volt power service lines that run over carnival attraction rides, require which of the following mandatory installation procedures:
    a. A clear space of no less than 15 feet in any direction
    b. Overcurrent protection at the ride location
    c. That each ride be separately connected to the power source
    d. None of the above

16. Equipment grounding terminal buses of essential branch-circuit panelboards that serve the same individual patient vicinity in a hospital must meet which of the following requirements:
    a. Be bonded together with a continuous, insulated conductor
    b. May not have conductors smaller than 10 AWG
    c. All of the above
    d. None of the above

17. A church sanctuary designed to hold 150 people is considered which of the following:
    a. An intrinsically safe location   b. An area of assembly
    c. A Class III location   d. None of the above

18. A construction trailer must have a minimum 100 amp electrical service.
    a. True   b. False

19. Non-metallic tubing cannot be installed above suspended ceilings in a location considered an assembly occupancy.
    a. True   b. False

20. Each patient bed in a critical care unit must have a minimum of six hospital-grade receptacles per location.
    a. True   b. False

21. The branch circuit conductor and overcurrent protection devices for an X-ray machine must be at least 50 percent of the momentary rating or 100 percent of the long-term rating, whichever is greater.

    a. True
    b. False

22. Requirements for patient beds located in critical care areas include which of the following:

    a. The location must have at least two branch circuits.
    b. The location must have one or more branch circuits from the emergency system and one or more from the normal power system.
    c. At least one of the branch circuits on the emergency system has to supply an outlet at each individual bed location.
    d. All of the above

23. Temporary electrical installations for holiday displays must be permanently removed after which of the following time periods:

    a. 31 days after installation
    b. 60 days after installation
    c. 90 days after installation
    d. 31 days after the end of the event

24. Any portion of a health care facility designed for patient examinations or treatment is considered to be which of the following:

    a. General care area
    b. Patient pare area
    c. Psychiatric treatment area
    d. None of the above

25. An ambulatory rehabilitation center that is limited to a total of 10 clients and that provides outpatient surgical treatment which includes only general anesthesia is not considered a health care facility.

    a. True
    b. False

Special Occupancies and Classifications

QUIZ 2—CLASSIFICATIONS

1. A Class I, Division 1 location is one in which ignitable concentrations of flammable gases or vapors could exist under unusual operating conditions.

   a. True   b. False

2. A Class III, Division 1 location is one in which easily ignitable fibers or materials that form combustible flyings, such as rayon and cotton, are handled, manufactured or used.

   a. True   b. False

3. Class II, Division 2 Group A pertains to locations that involve the use or production of methane.

   a. True   b. False

4. Class II Group E lists requirements for atmospheres that contain combustible metal dusts, including aluminum, magnesium, or others, with a particle size, abrasiveness, and conductivity that could present similar hazards in the use of electrical equipment.

   a. True   b. False

5. A Class I, Division 1 location is one where volatile flammable liquids or liquefied flammable gases are transferred from one container to another.

   a. True   b. False

6. In some Class 1 Division 1 locations, ignitable concentrations of flammable gases or vapor may be present continuously or for long periods of time, such as in the area between the inner and outer wall sections of a storage tank vehicle that contains volatile flammable fluids.

   a. True   b. False

7. Class III, Division 2 includes locations where easily ignitable fibers are handled during the process of manufacturing.

   a. True   b. False

8. A Class I, Division 1 location includes portions of cleaning and dyeing plants where flammable liquids are used.

   a. True   b. False

9. A Class I, Division 2 location is one in which ignitable concentrations of gases or vapors are normally prevented by positive mechanical ventilation, but that could become hazardous if there was a failure or abnormal operation of the ventilating equipment.

   a. True   **b.** False

10. In a Class II, Division 2 location, combustible dust could be present due to normal operations of equipment in quantities sufficient to produce explosive or ignitable mixtures.

    a. True   **b.** False

11. Class II Group G concerns atmospheres containing combustible dusts that are not included in Group E or F, and include such elements as flour, grain, wood, plastic, and chemicals.

    a. True   **b.** False

12. A typical Class I, Group C material is propane.

    a. True   **b.** False

13. Class II, Division 2 addresses areas where combustible dust accumulations that are on, in, or in the area of electrical equipment could be sufficient to interfere with the safe dispersal of heat from electrical equipment.

    a. True   **b.** False

14. Class I locations include those where flammable vapors are present in the air in quantities sufficient to produce ignitable mixtures.

    a. True   **b.** False

15. Class II locations are those in which ignitable concentrations of gases or vapors may often exist because of repair and maintenance operations or because of leakage.

    a. True   **b.** False

16. Ventilated areas within spraying or coating operations that use flammable liquids are considered hazardous, classified locations.

    a. True   **b.** False

17. Class II locations are those that are hazardous because of the presence of combustible dust.

    a. True   **b.** False

18. Dust coal, carbon black, charcoal, and coke dusts are examples of carbonaceous dusts.

    a. True   **b.** False

Special Occupancies and Classifications

19. Dust coal, carbon black, charcoal, and coke dusts are examples of carbonaceous dusts, which are outlined in Class II Group D locations.

    a. True   **b.** False

20. Class I, Group C locations include those where flammable gas, flammable liquid-produced vapor, or combustible liquid-produced vapor mixed with air that may burn or explode have either a maximum experimental safe gap (MESG) value greater than 0.45mm and less than or equal to 0.75mm, or a minimum igniting current ratio (MIC ratio) greater than 0.40 and less than or equal to 0.80.

    a. True   **b.** False

21. A typical Class I, Group C material is hydrogen.

    a. True   **b.** False

22. A typical Class I, Group B material is ethylene.

    a. True   **b.** False

23. Easily ignitable fibers and flyings include hemp, cocoa fiber, and Spanish moss.

    a. True   **b.** False

24. The interiors of refrigerators and freezers where volatile flammable materials are stored in open, lightly stoppered, or easily ruptured containers is an example of a Class I Division 2 location.

    a. True   **b.** False

25. Depending on factors such as quantity and size of the containers and ventilation, locations used for the storage of flammable liquids or liquefied or compressed gases in sealed containers may be considered either hazardous (classified) or unclassified locations.

    a. True   **b.** False

ANSWERS

QUIZ 1

1. B   Section [525.5(B)]
2. D   Section [517.19(A)]
3. A   Section [518.5]
4. B   Section [500.15(B)(2)(1)]
5. D   Section [517.30(B)(1)]
6. D   Section [551.81]
7. B   Section [500.5(B)(1) FPN 1]
8. A   Article 503
9. A   Section [513.7(F)]
10. C   Section [555.19(A)(4)]
11. A   Section [517.19(A)]
12. C   Section [550.31(1)]
13. C   Article 504
14. B   Section [511.12]
15. D   Section [525.5(B)] (NOTE: no rides may be placed were the power lines exceed 600 volts.)
16. C   Section [517.14]
17. B   Sections [518.1] and [518.2(A)]
18. B   Section [550.4(A)]
19. B   Section [518.4(C)(1) and (2)]
20. A   Section [517.19], Section [517.19(A)] and Section [517.19(B)(2)]
21. A   Section [517.73(A)]
22. D   Section [517.19], Section [517.19(A)] and Section [517.19(B)(2)]
23. C   Article 590
24. B   Section [517.2]
25. B   Section [517.1] Definition of Ambulatory Health Care Facility, requirement 2

QUIZ 2—CLASSIFICATIONS

1.	B	8.	A	14.	A	20.	A
2.	A	9.	A	15.	B	21.	B
3.	B	10.	B	16.	B	22.	B
4.	A	11.	A	17.	A	23.	A
5.	A	12.	B	18.	A	24.	B
6.	B	13.	A	19.	B	25.	A
7.	B						

—NOTES—

Chapter 6
SPECIAL EQUIPMENT

Electrician's Exam Study Guide

Chapter six of the NEC® outlines the requirements for special equipment, which, because of the nature of its use, construction, or unique characteristics, creates a need for additional safety measures that put it in a category of its own. These types of equipment include:

- Signs—Article 600
- Field manufactured wiring systems—Article 604
- Office partitions—Articles 605
- Cranes, hoists, elevators, dumbwaiters, wheelchair lifts, escalators, and other forms of lifting equipment—Articles 610 and 620.
- Electric vehicle charging systems—Article 625.
- Electric welders—Article 630.
- Amplifiers, computers, and other information technology devices—Articles 640 through 650.
- X-ray equipment—Article 660.
- Induction heaters, electrolytic cells, electroplating and irrigation machines—Articles 665 through 675.
- Swimming pools—Article 680.
- Integrated electrical systems—Article 685.
- "New fuel" technologies, such as solar-powered systems—Article 690.
- Fire pumps—Article 695.

Article 600 also covers neon and skeleton tubing, portable signs, and electrical art forms used for signage. Most commercial locations, such as stores in a shopping mall, have their own signs. Section **[600.5]** requires a sign outlet for the entrance of each tenant location, and Section **[600.5(C)(2)]** allows these signs to be used as junction or pull boxes for other adjacent signs and to contain both branch and secondary circuit conductors. There must be a disconnection means within the line of sight of a sign unless the disconnection means can be locked in the open position, per Section **[600.6(A)(1)]**. Section **[600.7]** tells us that sign installations have to be grounded, and you must use a minimum 14 AWG copper bonding conductors as required in Section **[600.7(D)]**. In Section **[600.9(A)]**, you'll learn that signs, like the ones you might install in a parking lot, must be located at least 14 feet above areas accessible to vehicles unless they're protected from physical damage.

> **! Code update SECTION [600.24]**
> This section was added to the 2005 NEC® and requires installations for outline lighting and signs supplied by Class 2 power sources to comply with the requirements of both Article 600 and Article 725.

Article 604 applies to field-installed manufactured wiring systems. These are subassemblies that have components that were included as part of the manufacturing process and therefore can't be inspected on-site without damaging the system. This article is very specific about cable and conduit requirements. For

Special Equipment | 141

example, Section **[604.6(A)(1)]** states that the only sizes allowed are Type AC cable and listed Type MC cable that has 600 volt, 8 to 12 AWG insulated copper conductors with either a bare or insulated copper equipment grounding conductor equal to the size of the ungrounded conductor. Conduits are specified in Section **[604.6(A)(2]** and can only be listed flexible metal or liquidtight types that contain the conductor styles and sizes listed in Section **[604.6(A)(1)].** There are two exceptions to these rules—one regarding luminaires, which cannot be any smaller than 18 AWG, and the other concerning conductors smaller than 12 AWG used in remote-control signaling and communication circuits.

Article 605 applies to relocatable wired office partitions, commonly known today as cubicles. Requirements for interconnections between partitions are listed in Section **[605.4(1–4)].** Lighting accessories and equipment installed as part of this type of office furnishings have their own compliance regulations, which are found in Section **[605.5(A–C)].** Cord-and-plug connections are allowed by Section **[605.8]**, as long as the total assembled partition or group of panels involved does not exceed 30 feet.

> **! Code** update SECTIONS [605.6] and [605.7]
>
> As a result of this change, these sections now require a disconnection means for a multiwire branch circuit that supplies either Fixed-Partitions [605.6] or Freestanding Partitions [605.7] to disconnect all ungrounded conductors at the same time. Disconnection means for multiwire branch circuits supplying these fixed and freestanding panels are now required to simultaneously disconnect all ungrounded conductors.

Moving away from the small items like moveable signs and relocatable walls, Article 610 outlines detailed requirements for the installation of electrical equipment and wiring connections for cranes, monorail and other hoists, and runways. Just as a reminder, if any of this equipment is located in a hazardous location, then the classifications and regulations of Articles 500, 501 and 503 apply. Article 600 is broken down into seven parts, as follows:

- Part I. General scope, and special requirements
- Part II. Wiring methods and conductors
- Part III. Contact conductors specific to runway crane bridges and tram-type tracks.
- Part IV. Disconnecting means
- Part V. Overcurrent protection
- Part VI. Controls
- Part VII. Grounding

Wiring methods for cranes and hoists are very specific and are listed in Section **[610.11(A–E)].** Here you will find that contact conductors and exposed conductors are not required to be enclosed in raceways. The allowable ampacities for insulated copper conductors, based on maximum operating temperatures, are listed in Table **[610.14(A)].**

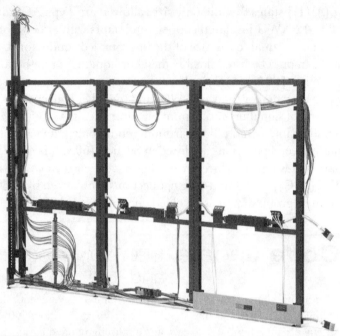

FIGURE 6.1 Wall partition connection.

> **! Code update SECTION [610.61]**
> This section now requires that any trolley or bridge frames of an overhead crane be bonded together and effectively grounded. The result is that bonding moving parts of the frame together minimizes shock hazard and eliminate stray voltages, eliminates the adverse effects of such external variants as paint, grease, dirt, or wear of the wheels and bearings, and eliminates the potential for ground loops.

Article 620 covers miscellaneous electrical equipment that carries or moves people, such as elevators, escalators, moving walks, wheelchair and stairway chair lifts, as well as one other form of moving equipment—dumbwaiters.

This Article is separated into the following parts:

- Part I. General scope, definitions of the elements associated with these kinds of equipment, such as elevator control rooms, and control systems, which are illustrated in FPN Figure **[620.2]**. Voltage limitations are also outlined.
- Part II. Conductors, including a Single-Line diagram in Figure **[620.13]**. Elevator feeder demand factors are listed based on the number of elevators on a single feeder in Table **[620.14]**.
- Part III. Wiring methods for the various equipment types, including branch-circuit requirements
- Part IV. Installation of conductors, including supports and wire fill requirements
- Part V. Traveling cable requirements
- Part VI. Disconnecting methods and controls, and identification requirements
- Part VII. Overcurrent protection standards

- Part VIII. Machine and control rooms and spaces
- Part IX. Grounding requirements
- Part X. Emergency and standby power systems

Article 625 is dedicated to an interesting category of electrical connections—Electric Vehicle Charging Systems. If you think this article consists of a few meager paragraphs, guess again. Section **[625.2]** defines an electric vehicle as an automotive "type" of vehicle, such as a passenger automobile, bus, truck or van, which is used for highway transportation and which has an electric motor that draws current from a rechargeable battery, fuel cell, or photovoltaic array as its power source. Part II, Section **[625.9(F)]**, requires that if an electric vehicle coupler is provided with a grounding pole, then it must be designed so that the pole connection is the first to make and the last to break contact.

> **! Code update SECTION [625.26]**
> This Section was added to cover new types of "Interactive charging systems" for hybrid vehicles that are either part of the vehicle (on-board) or separate from the vehicle. Now electric or hybrid vehicles will be able to supply power to buildings in the future.

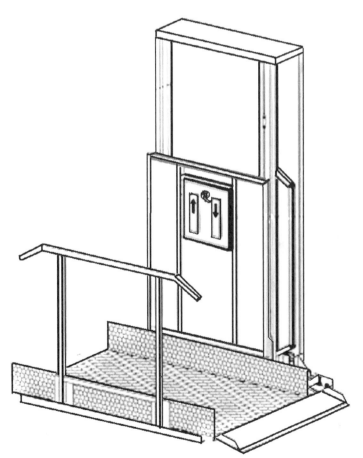

FIGURE 6.2 Wheelchair lift.

Because electric vehicles often contain chargeable batteries, they can release relatively high amounts of gas. This is why Part V, Section **[625.29(D)]**, requires adequate ventilation of these units. Section **[625.29(D)(2)]** lists calculation formulas for determining the minimum ventilation requirements for single and three-phase systems as illustrated below:

$$\text{Ventilation (single-phase) in cubic feet per minute (cfm)} = \frac{\text{(volts)(amperes)}}{48.7}$$

$$\text{Ventilation (three-phase) in cubic feet per minute (cfm)} = \frac{1.732\text{(volts)(amperes)}}{48.7}$$

> ▶ **Test** tip
>
> Remember that in the math chapter, we learned that three-phase systems are represented by 1.732.

Table **[625.29(D)(1)]** lists the minimum ventilation requirements in cubic meters per minute for the total number of electric vehicles that can be charged at one time, and Table **[625.29(D)(2)]** lists the same data in cubic feet per minute.

When you think about situations requiring circuit protection and current draw, remember that welding equipment either creates an electric arc between two surfaces, or it produces overcurrent that melts a welding rod. Article 630 focuses on adequately sizing the conductors and circuit protection to handle this type of load. After defining the general scope of this equipment in Part I, Article 630 breaks each type of welding equipment into its own Part as listed below:

- Part II. Arc welders
- Part III. Resistance welders
- Part IV. Welding cable

Section **[630.12(B)]** states that conductors that supply one or more welders must be protected by an overcurrent device that is rated at a maximum of 200 percent of the conductor ampacity. It includes the formula for calculating the maximum value of a rated supply current at its maximum rated output.

> **! Code** update SECTION [630.11]
>
> The 2005 NEC® changes the language concerning sizing conductors. The calculations have not changed, but the wording used in this section of the 2005 NEC® changed the calculated value of welding conductors to a "minimum size," instead of the required size. The size is based on 100 percent of the 2 largest welders, 85 percent of the 3rd largest welder, 70 percent of the 4th largest welder and 60 percent of the remaining welders.

Welding cable has several requirements that are different from other types of conductors. Section **[630.41]** states that the insulation on these cables must be flame-retardant and Section **[630.42(A)]** of the installation requirements states it must be supported at not less than 6-inch intervals.

Special Equipment

Article 640 is all about audio signal processing, amplification and reproduction equipment, such as recording equipment, public address systems, and permanently installed audio systems in places like restaurants, hotels and retail stores. Section **[640.3(A)]** states that abandoned audio cables must be removed. Section **[640.3(B)]** requires these types of installations to conform to the duct and plenum requirements in Section **[300.22]**. Article 640 specifies protection of audio equipment in wet locations, such as near bodies of water, in order to reduce the types of shock hazards particular to audio equipment installations. In addition, Article 640 makes a distinction between permanent and temporary audio installations. Part II provides requirements for permanent installations, and Part III provides requirements for temporary installations.

You might think that information technology (IT) equipment requires unique current loads and surge suppression and that this must be why it is covered in Article 645. Well, surprise. The main focus of this article is on preventing the spread of fire and smoke in IT areas and reducing the risk that fire or rescue workers might come in contact with energized equipment. Section **[645.10]** requires an IT shutoff switch to be readily accessible from the exit doors, so that there is a method to disconnect the power and battery sources before rescue personnel enter an IT room and start spraying fire hoses on energized equipment. A breakaway lock can protect the IT room from inadvertent shutdown through this switch.

Raised floor requirements, fire-resistant walls, separate HVAC systems, and other requirements further help achieve this goal.

> **! Code update SECTION [645.(D)(5)(C)]**
>
> Section [645.5(D)(5)(C)] now allows grounding conductors to be either green or green with one or more yellow stripes. The 2002 NEC® only allowed a green conductor to be used for grounding if it had one or more yellow stripes.

Article 647 provides information about avoiding power quality issues arising from the installation of sensitive electronic equipment, such as audio and visual systems. Ground loop is a common problem when multiple audio-visual system components are connected together. These ground loops often result in a humming noise in audio signals and interference bars in the image. Ground loop makes the system sensitive to interference from the main wiring installation and can lead to erratic operation of, or damage to, the equipments Article 647 defines proper wiring, grounding, and connection methods necessary to prevent ground loop and similar interference or equipment damage issues.

> **! Code update SECTION [647.4(A)]**
>
> This section change now permits the use of "double-pole fused disconnects" or double-pole switches and fuses to be used to simultaneously disconnect both ungrounded conductors from sensitive electronic circuits supplied by these panelboards.

Section **[647.7(A)(1)]** requires that all 15 amp and 20 amp receptacles must be GFCI protected, and Section **[647.7(A)(2)]** lists specific wording for any outlet strips, adapters and receptacle cover plates that may be used for connecting sensitive equipment, which should be:

WARNING—TECHNICAL POWER
Do not connect to lighting equipment.
For electronic equipment use only.
60/120 V. 1 phase ac
GFCI protected

Article 650 covers pipe organ installations and Article 660 lists the requirements for X-Ray equipment connections, conductor ratings, and overcurrent protection. Branch-circuit conductors and overcurrent protection device ampacity cannot be less than 50 percent of the momentary rating or 100 percent of the long-time rating of any X-ray equipment connected, per Section **[660.6(A)]**. The minimum size of 18 AWG or 16 AWG fixture wires is listed in Section **[660.9]** for control and operating circuits for X-ray equipment as long as it is protected by not larger than 20 amp overcurrent devices.

Articles 665 through 675 cover the topics of induction heaters, electrolytic cells, electroplating and irrigation machine installations and connections. Types of heating equipment for Article 665 are listed in Section **[665.2]**. Article 668 defines electrolytic cells as a tank or vat where electrochemical reactions are created by applying electrical energy in order to refine or create usable materials. Requirements and references for conductors, overcurrent protection, grounding, working zones and disconnection means are provide within the article. Industrial machinery is described as in the following code update:

> **! Code update SECTION [670.4 (B)–(C)]**
>
> The 2005 NEC® version of Section 670.4(B) clarifies the requirements for industrial machine disconnects by requiring that the machine is considered a "unit" and therefore requires a disconnect, that its branch circuit can be protected by either fuses or circuit breakers, and finally that the disconnect is not required to include any type of overcurrent protection as long as the protection is provided elsewhere. Section 670.4(C) lists the requirements for the overcurrent protection supplied to an industrial machine. They state that each supply circuit must be protected by a set of fuses or a single circuit breaker, that if the overcurrent protection is included in the machine, it must be marked to indicate that the overcurrent protection is at the machine's supply terminals, and finally that the supply conductors are either taps or feeders in accordance with Section 240.21.

Article 680 is dedicated to keeping people safe from the potential hazards posed by the electrical installations for swimming pools, fountains and similar installations. After defining the general requirements in Part I, Article 680 breaks out specific equipment in subsequent parts, which include permanently installed pools, storable pools, spas and hot tubs, fountains and pools, and tubs used for therapeutic purposes. Figure **[680.8]** illustrates the required clearances outlined in Table **[680.8]** for overhead conductors. Table **[680.10]** lists the minimum depth to which various wiring methods must adhere when buried. For example, if you are running wiring to a pool using rigid metal conduit, it must be at least 6 inches deep.

Article 690 lays out the requirements for solar photovoltaic installations and begins with Figure **[690.1]**, which identifies the various components of a solar power system. In the typical application, solar systems complement, rather than replace, a grid-based power source. For this reason, the very next illustration in Article 690 is Figure **[690.1(B)]**, which is a diagram of solar power components in common system configurations.

Special Equipment

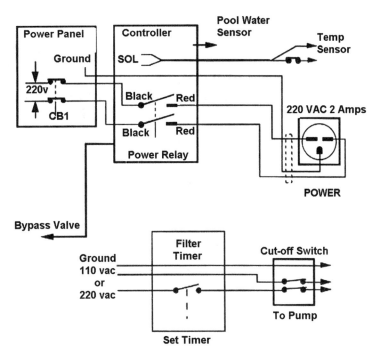

FIGURE 6.3 Pool wiring.

A number of cautionary warning requirements are listed in this article, as well as circuit sizing (Section **[690.8]**) overcurrent protection requirements (Section **[690.9]**) and specific additional provisions for this type of installation (Section **[690.14]**). A number of correction factors, based on the temperature rating of conductors, are listed in Table **[690.31(C)]**, and requirements regarding storage batteries in solar power systems are found in Part VIII, beginning with Section **[690.71].**

Unlike the disconnect methods and ideals behind most other types of electrical installations, a fire pump motor has to be able to run, no matter what, because it keeps water flowing to fire protection piping. Just keep this premise in mind as you review Article 695, especially when it seems that this article contradicts most of the rules for equipment up to this point in the NEC. For example, you would normally expect the disconnecting means on a power system to be lockable in the open position, but Section **[695.4(B)(2)]** specifies that a fire pump disconnect must be lockable in the closed position. Since a fire pump has to run without fail, its circuits can't have automatic protection against overloads. Section **[690.5(B)]** even requires that the primary overcurrent protection device be set to carry the sum of the locked-rotor current of the pump motor and the pressure maintenance pump motor and the full load current of any associated accessory equipment *indefinitely*. Secondary overcurrent protection is not permitted per this section.

> **! Code update SECTION [695.6(C)]**
>
> This Section of the 2005 NEC® requires that conductors that supply a fire pump be sized for 125 percent of the full load ampacity of the motor and must account for voltage drop.
>
> According to Section [695.7] the voltage drop at the motor terminals cannot be more than 5 percent below the voltage rating of the motor when the motor is operating at 115 percent of the full-load current rating of the motor.

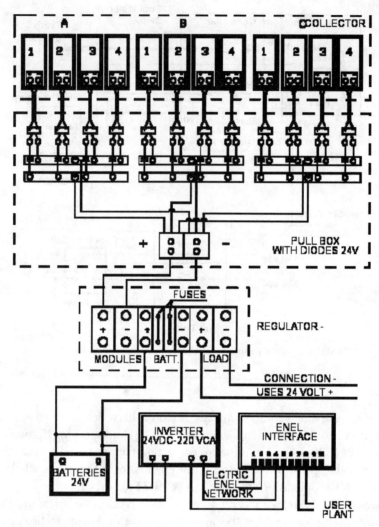

FIGURE 6.4 Photovoltaic diagram.

> ► **Test** tip
>
> Based on the requirements of Article 695, if the fire pump motor operates at 480 volts, a 5 percent voltage drop is 24 volts, or a minimum operating voltage of 456 volts. If the fire pump motor were operating at 208 volts, the maximum voltage drop is 14 volts, or a minimum operating voltage of 194 volts.

It is time once again to test your knowledge. The first part will be a quick quiz on requirements found in Article 600. There is no set time frame for this quiz.

Special Equipment   149

QUIZ 1

1. A therapeutic tub used for the submersion and treatment of patients shall not be which of the following:

    a. Protected by a ground-fault circuit interrupter

    b. Easily moved from one location to another

    c. Bonded

    d. All of the above

2. An electrical sign that uses incandescent lamps must be marked with the number of lampholders in the sign.

    a. True   **b.** False

3. The bottom edge of an electrical sign must be at least 12 feet above an area that is accessible to vehicles.

    a. True   **b.** False

4. Isolated parts that are no larger than 4 inches in any dimension and that are attached to a pool structure shall not require bonding if which of the following requirements is met:

    a. Connected to a no-niche luminaire bracket

    b. Listed for low-voltage

    c. They do not penetrate the pool structure more than 1 inch

    d. Accessible only by a qualified person

5. Type NM cables are allowed for use to connect data processing equipment in underfloor wiring installations.

    a. True   **b.** False

6. A means of disconnect must be provided for all data processing equipment in an information technology equipment room.

    a. True   **b.** False

7. In order to be considered storable, a swimming pool cannot exceed 12 feet in any dimension.

    a. True   **b.** False

8. Liquidtight flexible metal conduit for a hot tub installation shall be permitted if not more than which of the following lengths:
   a. 3 feet
   b. 5 feet
   c. 6 feet
   d. 10 feet

9. A disconnecting means must be provided for all HVAC equipment in an information technology equipment room.
   a. True
   b. False

10. A receptacle located within 20 feet of a swimming pool requires GFCI protection.
    a. True
    b. False

11. Unless listed otherwise, the minimum depth for an underwater lighting fixture is 18 inches below the water's surface.
    a. True
    b. False

12. The disconnection means in a data processing room must be near the principal exit door.
    a. True
    b. False

13. The walls of a bolted or welded metal pool may be used as the common bonding grid for this type of installation.
    a. True
    b. False

14. Any branch circuit that supplies underwater fixtures operating at over 15-volts must be protected by means of a GFCI.
    a. True
    b. False

15. A fountain that shares a water supply with a swimming pool does not have to comply with the installation requirements for the fountain and the swimming pool as well.
    a. True
    b. False

16. A steel junction box which is used to supply underwater fixtures does not also need to be a listed swimming pool junction box.
    a. True
    b. False

17. For a field-installed manufactured wiring system, Type AC cable that has 600 volt, 8 to 12 AWG insulated copper conductors with a bare grounding conductor equal to the size of the ungrounded conductor shall be allowed.
    a. True
    b. False

Special Equipment   151

18. The disconnecting means for a fire pump must be lockable in the closed position.

   a. True   **b.** False

19. A spa located inside the sun room in a single-family dwelling shall have at least one 125-volt receptacle that meets the following requirements:

   a. Is on a general-purpose branch circuit

   b. 15 amps or 20 amps

   c. Located not less than 5 feet from the inside wall of the spa

   d. All of the above

20. The required branch-circuit conductor ampacity for X-ray equipment must be at least 50 percent of the long-time rating of the equipment.

   a. True   **b.** False

21. Secondary overcurrent protection is required for fire pump wiring installations.

   a. True   **b.** False

22. Type MC cable with a 600 volt, 8 to 12 AWG insulated copper conductors and insulated copper equipment grounding conductor equal to the size of the ungrounded conductor is permitted for field-installed manufactured wiring system installations.

   a. True   **b.** False

23. Double-pole fused disconnects may be used in the installation of sensitive electronic equipment in order to create the required ground loop.

   a. True   **b.** False

24. Ungrounded conductors for multi-wire branch circuits that supply power to a freestanding office panel must be able to disconnect all ungrounded conductors separately.

   a. True   **b.** False

25. Where there is no fence or other barrier, a ground-fault circuit interrupter enclosure for a pool shall meet which of the following distance requirements:

   a. Be located not less than 6 inches vertically from the inside bottom of the box

   b. Be located not less than 6 inches vertically above the maximum water level

   c. Be located not less than 4 feet horizontally from the inside wall of the pool

   d. All of the above

26. Wiring for cranes and hoists enclosed in raceways which shall be permitted to be which of the following type:

   a. Type AC with insulated grounding conductors

   b. Contact conductors

   c. Short lengths of exposed conductors to resistors and collectors

   d. All of the above

27. An electrically interconnected assembly of electrolytic cells that is supplied by a direct-current power source is known as which of the following:

   a. Electrically connected cells

   b. Electrolytic cell system

   c. A cell line

   d. Intrinsically connected electrolytic units

28. Conductors that are external to crane or hoist motors and controls shall not be smaller than which of the following sizes:

   a. 20 AWG   **b.** 18 AWG

   c. 16 AWG   **d.** All of the above

29. What is the ampacity permitted for two hoist conductors in a raceway with an ambient temperature of 32 degrees C. and type USE #10 AWG for a short-term motor rated for 60 minutes:

   a. 30 amps   **b.** 37.6 amps

   c. 40 amps   **d.** 40.42 amps

30. The ampacity permitted for six hoist conductors in a cable with simultaneously energized 3 degrees C. AC power conductors to a short-term motor shall meet which of the following requirements:

   a. Each conductor shall be reduced by 80% of the value allowable for the four conductor rating

   b. 10 amps

   c. 26 amps

   d. None of the above

31. The maximum number of photovoltaic system disconnecting means permitted shall be not more than which of the following:

   a. 3 per circuit

   b. 6 switches or circuit breakers mounted in a single enclosure or switchboard

   c. 4 circuit breakers or fuses mounted in a group of separate enclosures

   d. 8

Special Equipment 153

32. Portable electric signs shall not be placed within which of the following locations of a fountain:

   a. In the fountain itself

   b. Within 10 feet of the fountain, measured horizontally from the inside wall

   c. Both of the above

   d. None of the above

33. Each resistive welder and control equipment shall be provided with which of the following:

   a. A temporary equipment connection

   b. A grounding conductor

   c. A switch or circuit breaker power disconnection means

   d. All of the above

34. Output from a stand-alone fuel cell system shall be permitted to supply which of the following:

   a. AC power to a building disconnecting means at levels of current below the rating of that disconnect means

   b. AC or DC power to a building disconnecting means

   c. AC power at an ampacity equal to that of the disconnecting means

   d. None of the above

35. Bridge wire contact conductors for cranes and hoists that span 85 feet shall meet which of the following requirements:

   a. Be kept at least 2 1/2 inches apart

   b. Shall have insulating saddles placed no farther apart than every 50 feet

   c. Both of the above

   d. None of the above

36. Any electrical connection on a marina dock for connections that are not intended to be submerged shall meet which of the following requirements:

   a. Be installed at least one foot above the dock decking

   b. Shall not be installed below the electrical datum plane

   c. Both of the above

   d. Either of the above

37. A single disconnecting means and all associated overcurrent protection shall be permitted for multiple remote fire power sources and which of the following:

    a. A listed fire pump power transfer switch

    b. An autotransformer

    c. A listed capacitor

    d. All of the above

38. If induction and dielectric heating equipment can be energized from multiple control points, then which of the following means shall be required:

    a. An interlocking means that ensures that the applicator can only be de-energized from one control point at a time

    b. An interlocking means that ensures that the applicator can be de-energized from any control point

    c. A limited power output of no more than 60 volts

    d. None of the above

39. Which of the following power supply sources shall be permitted to be installed between audio system equipment:

    a. Branch circuit power supplies   **b.** Transformer rectifiers

    c. Storage batteries   **d.** All of the above

40. If three crane motors are connected to the same branch circuit, then each tap conductor to each individual motor shall have an ampacity compliant with which of the following requirements:

    a. Not less than 1/3 the ampacity of the branch circuit

    b. An overload rating not more than the ampacity of the power circuit

    c. Be calculated using 110% of the of the sum of the crane motors

    d. None of the above

41. For an inside hot tub installation, which of the following parts shall be bonded together:

    a. All metal parts of electrical equipment that is associated with the hot tub circulation system, including pump motors

    b. All metal parts of electrical equipment that is associated with the hot tub circulation system, excluding pump motors

    c. Any metal surfaces located within 10 feet of the inside wall of the hot tub

    d. Electrical devices not associated with the hot tub but that are located not less than 6 feet from the hot tub

Special Equipment

42. Which of the following are permitted for use between hoistway risers and interlocks:
   a. Liquidtight flexible metal conduit
   b. Liquidtight flexible nonmetallic conduit
   c. Flexible metal conduit
   d. All of the above

43. The disconnecting means for individual electrically driven irrigation machine motors and controllers shall comply with which of the following:
   a. Be capable of simultaneously disconnecting all ungrounded conductors for each motor and controller
   b. Be readily accessible
   c. Shall not be located more than 75 feet from the machine location
   d. All of the above

44. The disconnect means for an elevator shall meet which of the following requirements:
   a. An enclosed fused motor circuit that is externally operable
   b. A circuit breaker capable of being locked in the closed position
   c. Capable of disconnecting the grounded power supply conductors
   d. None of the above

45. Whenever the electric connector for electric vehicle supply equipment is uncoupled from the vehicle there must be an interlock provided that does which of the following:
   a. Does not exceed 125 volts
   b. Does not exceed 20 amps
   c. De-energizes the connector
   d. All of the above

46. Which of the following equipment for an interior spa installation shall be grounded:
   a. All electric equipment used or associated with the spa circulation system
   b. Any electric equipment located within 8 feet of the inside wall of the spa
   c. All audio equipment
   d. All of the above

47. The maximum permitted distance between clamp-type intermediate supports for #4 AWG contact conductors is which of the following:
   a. 30 feet
   b. 9.0 m
   c. 60 feet
   d. 15 m

48. Branch circuits and feeders for electric vehicle supply equipment shall have overcurrent protection that is rated to comply with which of the following:

   a. Not less than 125% of the maximum load of the equipment

   b. Not more than 125% of the maximum load of the equipment

   c. Not less than 115% of the continuous load of the equipment

   d. Not more than 118% of the maximum load of the equipment

49. In a photovoltaic power system where a diversion charge controller is the sole means of regulating battery charging, which of the following shall be required:

   a. The conductor ampacity shall not exceed 150 percent of the power rating

   b. A utility-interactive inverter is required

   c. A second independent disconnect means that prevents overcharging of the battery

   d. None of the above

50. The area where conductive elements are placed on, in, or under a walk surface within 3 feet, that is bonded to metal structures and fixed nonmetallic equipment that could become energized, and that is connected to a grounding system in order to prevent voltage from developing within the area is known as which of the following:

   a. A forming shell

   b. The electrical datum plane

   c. An equipotential plane

   d. Strain relief zone

51. The minimum ampacity for supply conductors for a group of electric welders shall be based on which of the following:

   a. The sum of 100% of the two largest welders

   b. 85% of the third largest welder

   c. 70% of the fourth largest welder

   d. None of the above

52. Electric vehicle supply equipment cable shall be permitted to be which of the following types:

   a. Type EV or EVJ

   b. Type EVE or EVJE

   c. Type EVT or EVIT

   d. All of the above

53. Conductors that supply power for a group of three resistance welders shall have an overcurrent device that meets which of the following requirements:

   a. Can be disconnected from the power supply

   b. Is set at no more than 300 percent of the conductor ampacity

   c. Is set at no more than 200 percent of the rated primary current for each welder

   d. None of the above

Special Equipment

54. The input or output wires for an audio transformer shall not be connected directly to amplifier or loudspeaker terminals.

   a. True

   b. False

55. A disconnecting means shall be provided for electric vehicle supply equipment with which of the following ratings:

   I. Over 60 amps

   II. Over 150 volts to ground

   III. A continuous load of 100 amps or more

   IV. More than one vehicle connected to the equipment

   a. I & II b. II & III

   c. I & IV d. All of the above

56. Flexible metal conduit shall not be permitted for elevator car installations if which of the following conditions exists:

   a. Liquidtight flexible nonmetallic conduit can be used

   b. If the FMC is trade size 12 or larger

   c. If the length exceeds 6 feet

   d. All of the above

57. Audio system equipment shall not be placed in which of the following locations:

   a. Laterally within 5 feet of the inside wall of a pool

   b. Further than 5 feet from a power supply

   c. On a separately derived system

   d. None of the above

58. Any photovoltaic source and output circuit shall have ground-fault protection that complies with which of the following requirements:

   a. Detects a ground fault

   b. Automatically disconnects the conductors shutting off the utility-interactive inverter for that portion of the faulted array

   c. Includes a means to indicate that a ground fault has occurred

   d. All of the above

59. For outdoor portable use of audio equipment, which of the following shall be used to connect equipment racks and branch circuits:

   a. A listed, extra-hard usage cord

   b. Cables that are listed as suitable for wet location and sunlight-resistant use

   c. Both of the above

   d. None of the above

60. Photovoltaic modules shall be marked with identification of which of the following ratings:

   I. Operating voltage and current

   II. Open-circuit voltage

   III. Maximum allowable system voltage and maximum power

   IV. Short-circuit current

   a. I & III   **b.** II & III

   c. III   **d.** I, II, III & IV

61. If sensitive electronic equipment is supplied by 3-phase power, which of the following requirements shall be met:

   a. A separately derived 6-phase wye system shall be used

   b. The system shall have 60 volts to ground

   c. Shall be configured as 3 separately derived 120 volt single-phase systems with a total of no more than 6 main disconnects

   d. All of the above

62. Cords or cables for audio system equipment that are accessible to the public and are run on a floor or the ground shall be covered with which of the following:

   a. An approved non-conductive mat

   b. Black tape at least 4 inches wide

   c. Both of the above

   d. Either of the above

63. What is the ampacity permitted for four DC hoist conductors in a raceway with a maximum operating temperature of 257 degrees F. and type Z #12 AWG wire for a short-term motor rated for 30 minutes:

   a. 80 amps   **b.** 65 amps

   c. 50 amps   **d.** 45 amps

Special Equipment 159

64. Individual branch-circuits shall not be required for portable, transportable or mobile X-ray equipment that requires which of the following:

   a. A plug connection

   b. 60 amperes or lower capacity

   c. An operating interval of 5 minutes or less

   d. All of the above

65. Branch-circuit conductors that supply one or more data processing systems shall be of an ampacity meeting which of the following requirements:

   a. Is limited to 750 volt-amperes or less

   b. Is not less than 125% of the total connected load

   c. Is equal to the total connected load

   d. None of the above

66. Elevator machine rooms shall include which of the following:

   a. At least one 125 volt single-phase duplex receptacle

   b. An explosion-proof light switch

   c. Lighting that is connected to the load side of a ground fault circuit interrupter

   d. All of the above

67. When a separately derived 120 volt single-phase 3 wire system is used to reduce objectionable noise in sensitive equipment for a commercial occupancy, the system shall meet which of the following requirements:

   a. Include two ungrounded conductors, 60 volts each, to a grounded neutral conductor

   b. Not have a voltage drop in excess of 1.5 percent

   c. The combined voltage drop of the feeder and branch-circuit conductors does not exceed 2.5 percent

   d. All of the above

68. In order to determine the ampacity permitted for four hoists connected on a common #12 AWG conductor system, which of following is required:

   a. Apply a demand factor of 0.87

   b. Add together the permitted motor minimum ampacity for each hoist

   c. Multiply the sum of all of the allowable hoist motor ampacities by a demand factor of 0.87

   d. None of the above

69. All light fixtures connected to separately derived sensitive electronics systems operating at 60 volts to ground must have which of the following:

    a. A disconnect means located within sight of the luminaire and that can be locked in the open position
    b. A disconnection means that simultaneously opens all ungrounded conductors
    c. An exposed lamp screw-shell
    d. All of the above

70. The sum of the cross-sectional area of the individual conductors in a nonmetallic dumbwaiter wireway shall comply with which of the following requirements:

    a. Be no greater than 100% of the largest conductors
    b. Not be more than 50% of the interior cross-sectional area of the wireway
    c. Shall not exceed the maximum trade size permitted for the conduit
    d. None of the above

71. Overcurrent protection for pipe organ circuits shall meet which of the following requirements:

    a. Not be required for a common return conductor
    b. The 26 AWG conductor protection device is rated at no more than 6 amps
    c. The 28 AWG conductor protection device is rated at no more than 6 amps
    d. All of the above

72. X-ray equipment that is to be installed in an ambulance or other vehicle is considered to be which of the following:

    a. Portable
    b. Mobile
    c. Transportable
    d. Frequency-limited

73. Power receptacles for a pool circulation system shall not be installed in which of the following manners:

    a. Located at least 6 feet from the inside wall of the pool
    b. Located no less than 3 feet from the inside wall of the pool
    c. Consist of up to four receptacles
    d. All of the above

74. Single feeder runway supply conductors for a monorail do not require overcurrent devices.

    a. True
    b. False

Special Equipment

75. Which of the following shall be provided at the termination of a flexible cord of an underwater pool light fixture within a junction box:

    a. Strain relief
    b. A sealed hub
    c. A waterproof connection
    d. None of the above

76. The maximum photovoltaic source circuit current permitted shall be based on which of the following:

    a. The sum of the parallel module currents
    b. No greater than 600 volts
    c. The sum of parallel module rated short-circuit currents multiplied by 125 percent
    d. None of the above

77. If there are three AC crane conductors in a raceway with a maximum operating temperature of 257 degrees F, then the maximum size type PFA wire for a short-time rated motor for 30 minutes shall be which of the following:

    a. 20 AWG
    b. 18 AWG
    c. 16 AWG
    d. None of the above

78. Conductors and overcurrent devices for a feeder for two or more X-ray units shall have a rated ampacity of which of the following:

    a. Not less than 100% of the momentary demand rating of the two largest units
    b. Not less than 100% of the momentary demand rating of the two largest units plus 20 percent of the momentary rating of all other X-ray apparatus
    c. Not less than 100% of the momentary demand rating of the two largest units plus 20 percent of the long-time rating of all other X-ray apparatus
    d. Not less than 100% of the momentary demand rating of the two largest units and not more than 20 percent of the long-time rating of all other X-ray apparatus

79. The female half of an AC multiple pole branch circuit cable connector shall meet which of the following requirements:

    a. Shall be polarized
    b. Shall not be compatible with nonlocking connectors rated 250 volts or greater
    c. Be attached to the load end of the power supply
    d. All of the above

80. Contact conductors which are used as feeders for cranes or hoists shall meet which of the following requirements:

   a. Shall be designed to reduce sparks

   b. Shall not be used as feeders for any other equipment

   c. Shall not be guarded

   d. All of the above

81. Diffraction and irradiation-type equipment that is not effectively enclosed or provided with interlocks that prevent access to live current-carrying parts during operation shall have a positive indicating means when energized, which shall be which of the following approved methods:

   a. A pilot light or readable meter deflection

   b. A flashing red or yellow indicator

   c. An alarm or other suitable signaling device

   d. None of the above

82. Each unit of an information technology system that is supplied by a branch circuit shall include a manufacturer's nameplate with which of the following input power requirements:

   a. Voltage

   b. Frequency

   c. Maximum rated load amperes

   d. All of the above

83. The space envelope of an electrolytic cell line work zone includes which of the following spaces:

   a. Up to 96 inches above the energized surfaces of the cell line

   b. Within a 5 foot horizontal radius of energized surfaces of a cell line

   c. Areas below the energized surface of the cell line up to 96 inches, regardless of floors, barriers or partitions

   d. All of the above

84. A photovoltaic array with two outputs, each of which have opposite polarity to a common center tap, is considered to be which of the following:

   a. An array

   b. Bipolar array

   c. A blocking diode circuit

   d. A reverse array

85. Power supply conductors for fire pumps shall meet which of the following requirements:

   a. Be listed for wet locations

   b. Connect directly to a listed fire pump controller

   c. Be rated at 600 volts or more

   d. All of the above

Special Equipment

86. Power supply circuits to ungrounded receptacles for use with hand-held, cord-connected electrolytic cell equipment shall meet which of the following requirements:

   I. Be electrically isolated from the distribution systems that supply areas other than the cell line work zone

   II. Be grounded

   III. Have a power supply generated from an autotransformer

   IV. Include primaries that do not operate at more than 600 volts between conductors and that have overcurrent protection

   a. I & II
   b. II & III
   c. I & IV
   d. I, III & IV

87. A pushbutton shall be permitted as a disconnect means for both the information technology equipment room and the HVAC system serving that room, provided which of the following requirements is met:

   a. Pushing the button results in disconnecting the power
   b. The button is identified and readily accessible by the equipment room door
   c. The pushbutton is enclosed in a tamper-proof enclosure
   d. Never; a pushbutton is prohibited as a disconnect means for both systems.

88. For electrical components that supply DC power for electroplating and anodizing equipment, conductors shall be protected by which of the following overcurrent protection devices:

   a. Fuses and circuit breakers
   b. Removable links
   c. DC current arrestor
   d. None; DC power for this application does not require overcurrent protection.

89. For elevator cars, a separate branch circuit is required for which of the following power systems:

   a. Lights and auxiliary lighting
   b. Receptacles
   c. Ventilation
   d. All of the above

90. Supply conductors for industrial machinery shall have an ampacity no less than 125 percent of the full-load current of all resistance heating loads plus which of the following:

   a. 125 percent of the full-load current rating of the highest rated motor
   b. 125 percent of the full-load current rating of the highest rated motor plus the sum of the full-load current rating of all other connected motors based on duty cycles that may be in operation at the same time
   c. 125 percent of the total current ratings of all other connected motors based on duty cycles that may be in operation at the same time
   d. None of the above

91. Conductors that supply power for more than two welders shall meet which of the following requirements:

   a. Be no smaller than #12 AWG copper
   b. Be protected by an overcurrent device set at no more than 200 percent of the conductor ampacity
   c. Be provided with a disconnecting means
   d. All of the above

92. The supply circuit for X-ray equipment shall have a disconnecting means of adequate capacity for which of the following:

   a. At least 50% of the input required for the momentary equipment rating
   b. At least 100% of the input required for the long-time rating of the equipment
   c. The greater of either A or B
   d. Both A and B

93. Several irrigation motors shall be permitted on one branch circuit under which of the following conditions:

   a. The motors do not exceed a 2 horsepower rating each and each circuit has individual overload protection
   b. The full-load rating of any of the motors does not exceed 6 amperes
   c. Taps for each motor shall not be smaller than 14 AWG and no more than 15 feet in length
   d. All of the above

94. Hoistway door interlock wiring shall be which of the following:

   a. Type SF cable
   b. Type MTW
   c. Type TF
   d. All of the above

95. Supply circuit conductors for X-ray equipment shall have an ampacity that meets which of the following requirements:

   a. Is at least 50% of the input required for the momentary equipment rating
   b. Is at least 100% of the input required for the long-time rating of the equipment
   c. The greater of either A or B
   d. Both A and B

96. Conductors and optical fibers located in escalators or wheelchair lifts shall be installed in which of the following manners:

   a. In flexible metal conduit
   b. Be type EV cable
   c. In RMC
   d. None of the above

Special Equipment

97. The inverter output of a stand-alone photovoltaic system shall be permitted to supply which of the following:

   a. A 120 volt to single-phase 3 wire system

   b. 120 volt to single-phase 3 wire 120/240 volt distribution panel with no 240-volt outlets or multi-wire branch circuits

   c. 120 volt to single-phase 3 wire 120/240 volt distribution panel with 240-volt outlets and multi-wire branch circuits

   d. None of the above

98. The complete grouping of equipment used to convert chemical fuel into usable electricity and which typically consists of a reformer, stack, power inverter and auxiliary equipment is considered to be which of the following:

   a. An interactive system
   b. A fuel cell system
   c. A stand-alone power system
   d. A fuel regeneration system

99. A junction box that is connected to conduit that extends directly to a forming shell of a no-niche pool luminaire shall meet which of the following requirements:

   a. Be of a non-metallic material

   b. Be equipped with threaded hubs

   c. Be located at least 6 inches from the inside wall of the pool

   d. All of the above

100. In a fuel cell output circuit, which of the following shall be terminated at the grounded circuit conductor terminal of the premises wiring system:

   a. One conductor of a 2-wire system rated over 50 volts

   b. A neutral conductor of a 3-wire system

   c. Both of the above

   d. None of the above

QUIZ 2

1. A simultaneous disconnection means shall be required for multiwire branch circuits supplying fixed-type office furnishing partitions.
   - **a.** True
   - **b.** False

2. A junction box for a pool lighting installation shall not be located less than 4 feet from the inside wall of the pool, unless it is separated by a fence or wall.
   - **a.** True
   - **b.** False

3. In order to reduce the spread of fire and smoke, removal of abandoned audio cables is required.
   - **a.** True
   - **b.** False

4. Manufactured wiring systems are required to be secured only by use of staples or hangers.
   - **a.** True
   - **b.** False

5. Two- or three-pole circuit breakers, or single-pole circuit breakers with handle ties, shall be permitted to be used as a simultaneous disconnecting means for fixed office partitions.
   - **a.** True
   - **b.** False

6. Manufactured wiring systems of the flexible cord type shall be permitted to be used for electric-discharge luminaries.
   - **a.** True
   - **b.** False

7. Vehicle charging connectors shall be configured so that they are interchangeable with other electrical systems.
   - **a.** True
   - **b.** False

8. All metal fittings within or connected to a pool structure must be bonded.
   - **a.** True
   - **b.** False

9. A simultaneous disconnection means is not required for multiwire branch circuits supplying free-standing-type office furnishing partitions.
   - **a.** True
   - **b.** False

10. Vehicle charging connectors shall be configured to prevent unintentional disconnection:
    - **a.** True
    - **b.** False

Special Equipment 168

11. Motors associated with pool equipment shall be connected to an equipment grounding conductor that is not smaller than #14 AWG.

    a. True   **b.** False

12. A section sign is a sign or outline lighting system that is shipped as subassemblies and requires field-installed wiring between the subassemblies to complete the overall sign.

    a. True   **b.** False

13. An installation involving live parts, other than lamps, and neon tubing shall be enclosed.

    a. True   **b.** False

14. A field-installed secondary circuit wiring of section signs shall not exceed 1000 volts.

    a. True   **b.** False

15. Ground fault interrupters are not required in a branch circuit that supplies underwater pool lighting fixtures that operate over 15 volts.

    a. True   **b.** False

16. Transformers and power supplies provided with an integral enclosure, including a primary and secondary circuit splice enclosures, shall not require an additional enclosure.

    a. True   **b.** False

17. Radiant heating cables are permitted if embedded in or below the deck of a pool.

    a. True   **b.** False

18. Electrode receptacles shall be listed.

    a. True   **b.** False

19. Neon secondary-circuit conductors that are 1000 volts are permitted to run between the ends of neon tubing or to the secondary circuit midpoint return of listed transformers or listed electronic power supplies.

    a. True   **b.** False

20. Flexible cords for any type of pool shall not exceed 3 feet.

    a. True   **b.** False

21. Neon secondary-circuit conductors that are 800 volts shall be listed, insulated, and not smaller than #14 AWG.

    a. True   b. False

22. The location for a pool recirculating pump motor receptacle shall be no greater than 6 feet from the inside walls of the pool.

    a. True   b. False

23. All electrical equipment used for a storable swimming pool, including power supply cords, shall be protected by a GFCI.

    a. True   b. False

24. Metal conduit installed within 6 feet of the inside walls of a pool that are not separated from the pool by a permanent barrier shall be required to be bonded.

    a. True   b. False

25. The minimum spacing between neon tubing and the nearest surface shall be no less than 1/4 inch.

    a. True   b. False

26. Switching devices shall be located at least 5 feet horizontally from the walls of a pool.

    a. True   b. False

27. Unless sign lighting system equipment is protected from physical damage, the minimum installation height above vehicles shall be 16 feet.

    a. True   b. False

28. Swimming pool light fixtures shall not be installed on supply circuits that are over 150 volts between conductors.

    a. True   b. False

29. Only a separate HVAC system dedicated for information technology equipment use and separated from other areas of occupancy shall be permitted.

    a. True   b. False

30. The smallest permitted size for external conductors for crane motors and controls shall be 18 AWG.

    a. True   b. False

Special Equipment

31. Underground wiring is prohibited from running under a pool.
    a. True   **b.** False

32. Supply conductors shall directly connect a power source to a listed fire pump controller and power transfer switch.
    a. True   **b.** False

33. Photovoltaic output circuits are the conductors that run between the photovoltaic source circuits and the inverter or DC utilization equipment.
    a. True   **b.** False

34. A separate branch circuit is required for elevator door controls.
    a. True   **b.** False

35. Overcurrent devices set at no more than 200 percent of the conductor ampacity shall be required for conductors supplying power for more than two electric welders.
    a. True   **b.** False

36. Storage batteries shall not be installed as power supply sources between audio system equipment.
    a. True   **b.** False

37. Two-wire DC circuits for integrated electrical systems shall be permitted to be ungrounded.
    a. True   **b.** False

38. DC multipole connectors used to make a connection between loudspeakers shall only be compatible with nonlocking 15 or 20 amp connectors intended for branch circuit power.
    a. True   **b.** False

39. The output circuits of a solar photovoltaic AC module is considered an inverter output circuit.
    a. True   **b.** False

40. Partial shunting of electrolytic cell line circuits shall not be permitted.
    a. True   **b.** False

41. For photovoltaic systems, branch-circuit or supplementary overcurrent protection devices shall be permitted provide overcurrent protection.
    a. True   **b.** False

42. A single disconnecting means for both the electronic equipment and HVAC systems for an information technology equipment room is prohibited.

    a. True   **b.** False

43. Lead-acid storage batteries for photovoltaic systems for dwelling units shall not have more twenty-four 2-volt cells connected in series.

    a. True   **b.** False

44. A light fixture designed to be installed above or below the water level in a fountain without a niche is termed a dry-niche luminaire.

    a. True   **b.** False

45. Permanently wired sensitive electronic utilization equipment must have an equipment grounding conductor with circuit conductors to an equipment grounding bus that is permanently labeled the "Technical Equipment Ground" at the panelboard where the branch-circuit originates.

    a. True   **b.** False

46. A main common-return conductor in the electromagnetic supply for pipe organs shall not be less than 14 AWG:

    a. True   **b.** False

47. A separately derived 120 volt single-phase 3 wire system is required to reduce objectionable noise in sensitive equipment installed in a multifamily or commercial occupancy.

    a. True   **b.** False

48. Power and communication cables for information technology equipment shall be permitted under raised floors provided that a ventilation system in the underfloor area exists with approved smoke detectors, and that will cease air flow circulation upon the detection of fire or combustion.

    a. True   **b.** False

49. A light fixture designed to be installed in a niche that is sealed against the entry of any pool water is considered to be a wet-niche luminaire:

    a. True   **b.** False

50. Conductors for a feeder for three X-ray units shall have a rated ampacity that is not less than 100% of the momentary demand rating of the two largest units plus 20% of the momentary rating of the third X-ray unit.

    a. True   **b.** False

Special Equipment | **171**

51. A bonding jumper shall be installed on the grounding electrode conductor for a photovoltaic system.
   a. True **b.** False

52. DC cell line process power-supply conductors shall be grounded.
   a. True **b.** False

53. Forming shells shall be installed for all wet-niche underwater pool light mounting.
   a. True **b.** False

54. Overcurrent devices set at no more than 300 percent of the conductor ampacity shall be required for conductors supplying power to a group of resistance welders.
   a. True **b.** False

55. In a solar photovoltaic array, open circuiting, short circuiting or opaque covering shall be used to disable the array.
   a. True **b.** False

56. AC cell systems that supply portable electrical equipment within a cell line working zone shall be grounded:
   a. True **b.** False

57. Photovoltaic source circuit isolating switches, blocking diodes and overcurrent devices shall not be permitted on the photovoltaic side of photovoltaic disconnecting means.
   a. True **b.** False

58. Branch circuit conductors and overcurrent protection devices for electrically driven irrigation machines shall have an equivalent continuous-current rating equal to 125 percent of the machinery motor nameplate full-load current rating of the largest motor plus a quantity equal to the sum of each motor full-load current rating for any remaining motors on the circuit, multiplied by the maximum percent duty cycle at which the motors can continuously operate.
   a. True **b.** False

59. An AC module system for a solar photovoltaic system includes an inverter and dedicated branch circuits for the electric production and distribution network.
   a. True **b.** False

60. Any 125-volt receptacles located within 20 feet of the inside walls of a storable pool must be protected by ground-fault circuit interrupters.
   a. True **b.** False

61. Audio equipment installations that could present a risk of electric shock or fire to the public shall be protected by a barrier or supervised by a qualified person.

    a. True   **b.** False

62. An inverter is equipment used to change voltage levels or waveforms, typically AC input to DC output.

    a. True   **b.** False

63. A photovoltaic source circuit provides DC power at system voltage.

    a. True   **b.** False

64. Conductors for electronic signal circuits for pipe organs shall not exceed 28 AWG.

    a. True   **b.** False

65. All 15 amp and 20 amp receptacles used to connect sensitive electronic equipment shall be GFCI protected.

    a. True   **b.** False

66. The photovoltaic source and output circuits for single-family dwelling fixtures shall be permitted to have a maximum photovoltaic system voltage of up to 600 volts as long as the circuits include at least one fixture or receptacle.

    a. True   **b.** False

67. An interlock that de-energizes the electric connector for electric vehicle supply equipment when it is uncoupled from the vehicle is required.

    a. True   **b.** False

68. Securely fastened cabling shall be permitted to be attached directly to a pipe organ structure without insulating supports.

    a. True   **b.** False

69. Mobile X-ray equipment of any ampere capacity shall be supplied through a suitable hard-service cable or cord.

    a. True   **b.** False

70. On a dock, any electrical connections that are not intended to be submerged shall be 24 feet above the decking.

    a. True   **b.** False

Special Equipment

71. Series-connected strings of two or more photovoltaic system modules shall have individual overcurrent protection devices.

    a. True   **b.** False

72. A fuel cell system transfer switch in a non-grid-interactive system that uses utility grid back-up shall not be required.

    a. True   **b.** False

73. Contact conductors used as feeders for hoists shall not be used as feeders for any other equipment.

    a. True   **b.** False

74. A forming shell is a structure designed to enclose a luminaire that is installed in a pool or fountain.

    a. True   **b.** False

75. Conductors for electromagnetic valve supply circuits for pipe organs shall not be less than 26 AWG.

    a. True   **b.** False

ANSWERS

QUIZ 1

1. **B** Section 680.62
2. **A** Section 600.4(B)
3. **B** Section 600.9(A) must be at least 14 feet above vehicles
4. **C** Section 680.26(3)
5. **B** Section 645.5(D)(2)
6. **A** Section 645.10
7. **B** Section 680.2
8. **C** Section 680.42(A)
9. **A** Section 645.10
10. **A** Section 680.22(A)(5)
11. **B** Section 680.23(A)(5)
12. **A** Section 645.10
13. **A** Section 680.26(C)(2)
14. **A** Section 680.23(A)(3)
15. **B** Section 680.50
16. **B** Section 680.24(A)(1)
17. **A** Section 604.6(A)(1)
18. **A** Section 695.4(B)(2)
19. **D** Section 680.43(A)
20. **B** Section 660.6(A)
21. **B** Section 695.4(B)(2)
22. **A** Section 604.6(A)(1)
23. **B** Section 647.4(A)

> ► **Test** tip
> The goal of this Section of the code is to prevent ground loops.

24. **B** Sections 605.6 and 605.7
25. **C** Section 680.24(B)(2)(b)
26. **A** Section 610.11(A) and (B)
27. **C** Section 668.2 Definitions
28. **C** Section 610.14(C) (NOTE: the question does not provide or allow for any of the exceptions listed in C (1) or (2).)

Special Equipment

29.	B	Table 610.14(A) with Ampacity Correction Factors applied	65.	B	Section 645.5(A)
30.	A	Table 610.14(A) Footnote 2	66.	A	Section 620.23(A–C)
31.	B	Section 690.14(C)(4)	67.	D	Section 647.3 to 647.4(D)
32.	A	Section 680.57(C)(2)	68.	C	610.14(E)(3)
33.	C	Section 630.33	69.	B	Section 647.8(A–C)
34.	A	Section 692.10(A)	70.	B	Section 620.32
35.	C	Section 610.21(E)	71.	D	Section 650.8
36.	C	Section 682.12	72.	C	Section 660.2 Definitions
37.	A	Section 695.4(B)(1)	73.	D	Section 682.22(A)(1)
38.	A	Section 665.7(A)	74.	B	Section 610.41(A)
39.	D	Section 640.21(D)(1–4)	75.	A	Section 680.24(E)
40.	A	Section 610.42(B)(1)	76.	C	Section 690.8(A)(1)
41.	A	Section 680.43(D)	77.	D	Table 610.14(A)
42.	D	Section 620.21(A)(1)	78.	C	Section 660.6(B)
43.	A	Section 675.8(C)	79.	D	Section 640.41
44.	A	Section 620.51	80.	B	Section 610.21(H)
45.	C	Section 625.18	81.	A	Section 660.23(B)
46.	A	Section 680.43(F)	82.	D	Section 645.16
47.	C	Table 610.14(D)	83.	A	Section 668.10(A)(1–3)
48.	A	Section 625.21	84.	B	Section 690.2 Definitions
49.	C	Section 690.72(B)(1)	85.	B	Section 695.4(A)
50.	C	Section 682.2 Definition	86.	C	Section 668.21(A)
51.	D	Section 630.11(B)	87.	A	Section 645-10
52.	D	Section 625.17	88.	A	Section 669.9
53.	B	Section 630.32(B)	89.	D	Section 620.22(A)
54.	B	640.9(D)	90.	B	Section 670.4
55.	A	Section 625.23	91.	B	Section 630.12(B)
56.	C	Section 620.12(A)(2)	92.	C	Section 660.5
57.	A	Section 640.10(A)	93.	D	Section 675.10
58.	D	Section 690.35(D)	94.	A	Section 620.11(A)
59.	C	Section 640.42(E)	95.	C	Section 660.6(A)
60.	D	Section 690.51	96.	C	Section 620.21
61.	D	Section 647.5	97.	B	Section 690.10(C)
62.	A	Section 640.45	98.	B	Section 692.2 Definitions
63.	C	Table 610.14(A)	99.	B	Section 680.24(A)(1)
64.	B	Section 660.4(B)	100.	C	Section 692.41(B)(1 and 2)

QUIZ 2

1. A	20. B	39. A	58. A
2. A	21. B	40. B	59. A
3. A	22. B	41. A	60. A
4. B	23. A	42. B	61. A
5. A	24. B	43. A	62. B
6. A	25. A	44. B	63. B
7. B	26. A	45. A	64. B
8. A	27. B	46. A	65. A
9. B	28. A	47. B	66. B
10. A	29. B	48. A	67. A
11. B	30. B	49. B	68. A
12. A	31. A	50. A	69. A
13. A	32. A	51. B	70. B
14. B	33. A	52. B	71. B
15. B	34. B	53. A	72. B
16. A	35. A	54. A	73. A
17. B	36. B	55. A	74. B
18. A	37. A	56. B	75. A
19. A	38. B	57. B	

Chapter 7
SPECIAL CONDITIONS

Electrician's Exam Study Guide

Special conditions warrant their own chapter in the NEC. Chapter seven is dedicated to such situations as when emergency power systems become activated, and addresses hospital communications and other legally required systems, stand-by systems designed to protect against economic loss or disruption, fire alarm systems, and fiber optic installations. These unique scenarios are covered in the following order:

- Emergency systems—Article 700
- Legally required standby systems—Article 701
- Optional standby systems—Article 702
- Interconnected power production systems—Article 705
- Circuits and equipment that run at less than 50 volts, and remote-control, signaling or power-limited circuits by classification—Articles 720 and 725
- Instrument tray cables for control circuits of 150 volts or less and 5 amperes or less—Article 727
- Fire alarm systems and requirements—Article 760
- Fiber optic cables, raceways and installations—Article 770

When was the last time you went to the movies? It's dark in there, isn't it? If the complex lost power or if the popcorn machine caught the place on fire, an emergency system would kick in to provide a lighted exit path. It is not a system designed for continuous normal operations, but to provide lighting and controls essential for safety in the event of (yes, you guessed it) an emergency. The requirements for the circuits and equipment used to control and distribute lights and/or power under these types of situations are covered in Article 700. An overall description of these applications is outlined in Section **[700.1 FPN: #3]**. This article does not just cover lighting, but also includes safe actuation devices, like a vent or valve that goes into a specific position if there is a loss of power. One other interesting aspect of this article is that it specifically states that the systems are legally required by the agency having jurisdiction.

FIGURE 7.1 Emergency illumination of exits is provided by clearly marked exit signs.

> **! Code update SECTION 700.9 (D) (1)**
> In the current code, feeder circuits for emergency systems must be installed in areas that are protected by an approved automatic fire suppression system. The 2002 Code said feeder circuits were required to be installed in *buildings that were protected* by an approved automatic fire suppression system. Based on the old wording, a feeder for an emergency system could be run above a sprinkler-covered space. The new wording makes it clear that emergency feeders can only be run in spaces that are protected by sprinklers, not just anywhere in a building that has sprinklers.

So if emergency systems in Article 700 are legally required, and Article 701 addresses legally required standby systems, then what is the difference? If you read the description of an emergency system in the second paragraph of Section **[701.1]** and compare it to the definition of a legally required standby system in Section **[701.2]**, you'll find they match word for word until the point where they describe what the systems are intended to do. Emergency systems automatically supply *illumination*, power, or both to specific areas and equipment, but legally required systems automatically supply *power only* and to selected loads. Additionally, legally required systems have to be able to supply standby power *in 60 seconds* or less after a power loss, instead of the *10 seconds* or less as specified for emergency systems. So, legally required systems would serve the power needs of such equipment or systems as a smoke removal system, a refrigeration system, or an industrial process that could create a hazard or hamper rescue efforts if it stopped because of a power loss. Here is another way to look at it: whereas Article 700 supplies lighting that provides an exit path, Article 701 lighting provides illumination of fire hydrants or switchgear areas that would be used during an emergency.

Article 702 explains optional standby systems that protect facilities or property in situations where these systems don't affect rescue operations. These systems are designed to protect against property or economic loss. Imagine that the data control center of one of the nation's largest banks lost power, the servers were not shut down properly, and millions of records were lost. Even worse, what if one of the major pharmaceutical centers lost power without a standby power system in place, and as a result their entire supply of rubella or flu vaccines was ruined? It can take over a year to culture vaccines. Article 702 may be short, but it is obviously important. The most important part as far as your passing the licensing exam goes is that you know what conditions are covered by optional standby systems. Section **[702.5]** specifies that these systems must have adequate capacity and ratings to supply all of the equipment that needs to be used all at one time. This means it has to carry the maximum fault current at all of its terminals.

> **▶ Test tip**
> Section [702.9] allows optional standby system wiring to occupy the same raceways, cables, boxes, and cabinets as other general wiring.

The requirements for electric power sources that operate in parallel with a primary source of electricity, such as a utility supply, are described in Article 705. Section **[705.21]** does not require disconnection means for equipment that is intended to operate as an integral part of a power production source that exceeds 1000 volts. If there is a loss of primary power, then Section **[705.40]** requires that the electrical power source be automatically disconnected from the entire primary source ungrounded conductors, and that it stay unconnected until primary power is restored.

Circuits that fall under Article 725 are remote control, signaling, and power-limited circuits that are not an integral part of a device or appliance. Article 725 includes circuits for burglar alarms, access control, sound, nurse call, intercoms, some computer networks, some lighting dimmer controls and some low-voltage industrial controls. Some examples of these types of circuits are:

- Remote control circuits that control other circuits through a relay or solid-state device, such as a motion-activated security lighting circuit
- A signaling circuit that provides output that is a signal or indicator, such as a buzzer, flashing light or annunciator
- A power-limited circuit that operates at a maximum of 30V and 1,000VA

Circuits are divided into Class 1, Class 2 and Class 3 types. Because the risk of fire is lower with qualifying circuits, Article 725 specifies conditions in which standards established in first few chapters of the NEC® are not required. For example, you don't have to put a Class 2 splice in a box or enclosure.

> **! Code update SECTION 725.56(F)**
> The 2005 edition added item #F to Section [725.56]. Audio amplifier circuits can use Class 2 or Class 3 wiring between the amplifier and audio speaker and are considered equivalent to Class 2 or Class 3 circuits. They are required to be insulated and installed like Class 2 or Class 3 circuits. Section [(F)] now makes it clear that these audio amplifier circuits cannot be installed in the same raceway as other Class 2 or Class 3 wiring.
> NOTE: If the audio circuit shorted out and was run in the same raceway as, say, the fire doors or smoke control systems, that short could disable those safety systems.

Circuits described in Article 725 are characterized by usage and power limitations that differentiate them from electric light and power circuits. The article provides alternative requirements for minimum wire sizes, derating factors, overcurrent protection, insulation requirements, wiring methods and materials.

> **! Code update SECTION 725.3(C)**
> Plenum signaling raceways are now permitted in spaces used for environmental air. It was commonly used for fiber optic cabling and communications wiring, but is now allowed for remote control, signaling and power-limited circuits per the 2005 code.

What would a code book designed to address safety issues be without an article focused on fire alarm systems? Article 760 provides the requirements for the installation of wiring and equipment for fire alarm systems, including all circuits the fire alarm system controls and powers. Fire alarm systems include fire detection and alarm notification, voice communications, guard's tour, sprinkler water flow, and sprinkler supervisory systems. As you study this article, be sure to focus your attention to the illustrations, which highlight important requirements that you may otherwise have difficulty understanding from the Article 760 text alone.

Special Conditions | 181

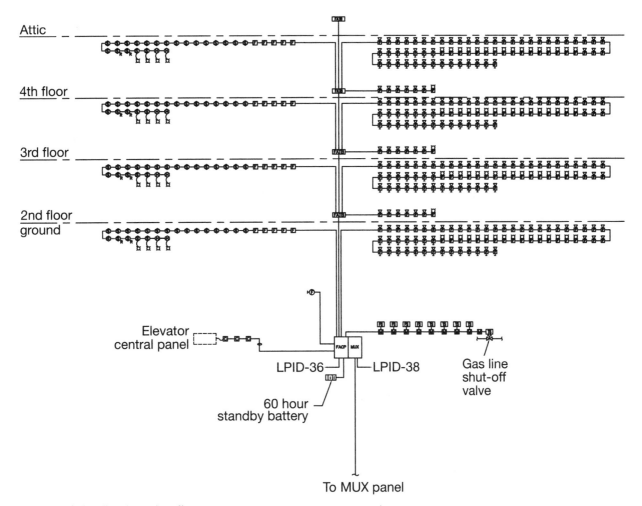

FIGURE 7.2 Fire alarm riser diagram.

Section **[760.21]** addresses non-power limited fire alarm (NPLFA) systems that have a voltage less than 600 volts nominal, and circuits that are not supplied through GFCI or arc-fault interrupters. Overcurrent protection for NPLFA conductors cannot exceed 7 amperes for 18 AWG conductors and 10 amperes for 16 AWG conductors. Conductors and circuits are divided into several classifications: Class 1 conductors of different circuits in the same raceway or cable enclosure (Section **[760.26]**), conductors with more than two different power limited fire alarm (PLFA) circuits, Class 2 and Class 3 conductors (Section **[760.56]**) and communications circuits that run in the same raceway. Figure **[760.61]** illustrates cable substitution allowances, and these are listed in Table **[760.1]**. The various types of cables used for fire alarm installations are listed, beginning with Section **[760.81]** and running through Section **[760.82(J)]**.

> ► **Test** tip
> Fire Alarm Circuit Integrity (CI) Cable versus Electrical Circuit Protective System: A fire alarm circuit integrity cable (CI) has a 2-hour fire resistance rating *without being installed in conduit*. An electrical circuit protective system is a cable with a 2-hour fire rating *installed in conduit*.

Unlike regular telephone lines, fiber-optic cable has the communication advantage of being highly scalable within the same infrastructure and eliminating electromagnetic noise. Article 770 provides the requirements for installing optical fiber cables and the optical raceways that contain and support them. It also contains the requirements for hybrid cabling that combines optical fibers with current-carrying metallic conductors. Section **[770.3]** references installation requirements found in Sections **[300.21]** and **[300.22].** It also addresses requirements for stopping the spread of combustion and installation methods for optical fiber raceways in ducts, plena or other air-handling spaces. On the other hand, Article 770 does not refer to back to Section **[300.15]**, which means you don't have to use boxes for splices or termination of optical fiber cable.

Since the NEC® is not a design guide and does not guarantee the most efficient installation processes, there are a number of technical fiber optic installation processes that are not addressed in Article 770. Don't bother to look for information on bend radius details or information on how to test out fiber optic systems. These are the kinds of techniques you have to go to BISCI trainings to know.

The final article in Chapter 7 of the NEC® is Article 780, which is provided for closed-loop and programmed power distribution systems. These are on-site power distribution systems that are jointly controlled by signaling between energy controlling equipment and utilization equipment. One of the key issues in this article is not so much energizing parts, but rather de-energizing connections. For example, Section **[780.3(B)]** states that the outlets in these system installations have to be de-energized when a ground-fault or overcurrent condition exists or if the nominal operation acknowledgement signal is not being received from equipment connected to the outlet. If an alternate power source is used, then Section **[780.3(C)]** indicates that outlets must be de-energized if a situation occurs in which the grounded conductor is not grounded properly or an ungrounded conductor is not at nominal voltage. Article 780 creates a smooth transition into the next NEC® chapter, which is all about communications systems.

Now that you have reviewed the various special condition requirements of Chapter seven, it is time once more to check your knowledge and see how you are doing with your test-taking skills. The quick quiz below is not a timed test. Full-blown practice exams are coming up soon, but for now, your goal is still to improve your ability to get as many questions correct as you can.

Special Conditions

QUIZ 1

1. Overcurrent protection for 18 AWG NPLFA conductors cannot exceed which of the following ratings:

   a. 7 amperes
   b. 10 amperes
   c. 15 amperes
   d. 20 amperes

2. Procedures for testing fiber optic equipment are established by the NEC.

   a. True
   b. False

3. A Class 1 power-limited circuit must be supplied by a power source that has a maximum rated output of which of the following:

   a. 30 volts 1000 amps
   b. 1000 amps
   c. Both of the above
   d. None of the above

4. If a normal power system fails, then the maximum delay allowed in a legally required emergency stand-by system before emergency power cuts on is which of the following:

   a. 10 seconds
   b. 30 seconds
   c. 60 seconds
   d. 2 minutes

5. A non-power limited fire alarm circuit cannot exceed which of the following:

   a. 150 volts
   b. 220 volts
   c. 240 volts
   d. 600 volts

6. When low-voltage fire alarm system cables are installed in a fire-resistant rated wall, which of the following are required:

   a. Cables may not penetrate through firewalls
   b. The cables must run through rigid conduit
   c. Openings must be firestopped
   d. There are no special considerations for low-voltage fire alarm cables

7. A power system designed to prevent loss of power that could cause economic loss or business interruptions is designated as which of the following:

   a. Legally required stand-by system
   b. Optional stand-by system
   c. Interconnected power stand-by system
   d. None of the above

8. An emergency stand-by system must be designed to provide power within 10 seconds of a primary power loss.
   a. True
   b. False

9. Provided that an alternate power source is adequately rated, it can be used to carry emergency loads.
   a. True
   b. False

10. To warn emergency response personnel, which of the following must be used to indicate the type and location of on-site emergency power sources:
    a. Diagram inside the panelboard cover
    b. A red light
    c. A sign
    d. A diagram posted in a central location, such as the main entrance of the building

11. Emergency circuits cannot supply loads other than those required for emergency use.
    a. True
    b. False

12. In the alternate power source for emergency systems, which of the following are true:
    a. Ground-fault protection does not have to be provided for equipment
    b. Ground-fault indication must be provided
    c. Both of the above
    d. None of the above

13. The branch-circuit wiring that supplies ELBE cannot be the same branch-circuit wiring that supplies the normal lighting in the area.
    a. True
    b. False

14. An emergency system power source must have adequate capacity to do which of the following:
    a. Safely carry emergency loads independently of each other
    b. Provide emergency power for up to 10 hours
    c. Safely carry all emergency loads expected to operate simultaneously
    d. None of the above

15. Type NPLFA conductors must be stranded copper of which of the following sizes:
    a. 14 AWG
    b. 16 AWG
    c. 18 AWG
    d. Either 14 AWG or 16 AWG

Special Conditions

16. Multiconductor non-power limited fire alarm cables must conform to which of the following requirements:

    a. Be marked with a maximum voltage usage rating of 150 volts

    b. Be capable of carrying a minimum capacity of 120 volts

    c. Both of the above

    d. None of the above

17. NPLFA circuits and Class 1 circuit conductors that run through a raceway may not carry a continuous load in excess of 50 percent of the ampacity of each conductor.

    a. True   b. False

18. Low-power network-powered broadband communications circuits are permitted in the same raceway as which of the following:

    a. Non-power limited cabling

    b. Power-limited fire alarm circuits

    c. Normal power cabling

    d. Low-power network-powered broadband circuits cannot run in raceways with any other type of circuits.

19. Instrumentation tray cable may not be installed on circuits that operate in excess of which of the following:

    a. 150 volts or 5 amps   b. 120 volts or 5 amps
    c. 100 volts or 10 amps   d. 115 volts or 5 amps

20. Overcurrent protection for NPLFA conductors cannot exceed 10 amperes for which of the following conductor sizes:

    a. 14 AWG   b. 16 AWG
    c. 18 AWG   d. None of the above

QUIZ 2

1. Peak load-shaving operation shall not be permitted as a means of testing emergency systems.
   - **a.** True
   - **b.** False

2. Portable emergency units shall be permitted.
   - **a.** True
   - **b.** False

3. Branch circuit overcurrent devices for legally required standby circuits shall be accessible only to authorized personnel.
   - **a.** True
   - **b.** False

4. Type OFCP nonconductive and conductive optical fiber cables are approved for use in ducts and plena.
   - **a.** True
   - **b.** False

5. On a wiring system that operates at 50 volts or less, a receptacle for a kitchen toaster shall be not less than 20 amps.
   - **a.** True
   - **b.** False

6. Transfer equipment shall be permitted to supply both the emergency and standard power systems.
   - **a.** True
   - **b.** False

7. The maximum setting for signal devices installed for an emergency system shall be 1000 amperes of less.
   - **a.** True
   - **b.** False

8. Feeder circuit wiring for emergency systems shall not be installed in areas that are not protected by a fire-rated assembly with a 1-hour fire rating.
   - **a.** True
   - **b.** False

9. If an internal combustion engine is used as the prime mover for a legally required standby system, then there shall be an on-site fuel supply capable of powering the system for no less than 1 1/2 hours at full-demand load.
   - **a.** True
   - **b.** False

Special Conditions

10. Feeder circuit wiring for emergency systems shall be installed in areas that are completely protected by an approved automatic fire suppression system.

    a. True   b. False

11. A means to protect against an emergency system automatically transferring between dual fuel supplies shall be provided.

    a. True   b. False

12. Individual emergency unit equipment shall include a battery that can maintain no less than 60 percent of the initial emergency illumination for at least 1 1/2 hours.

    a. True   b. False

13. Emergency circuits shall be permitted to supply appliances and lamps necessary to continue standard work practices without interrupting power in office buildings.

    a. True   b. False

14. The bypass isolation switches for legally required systems shall ensure parallel operation of the system.

    a. True   b. False

15. For emergency systems, a means of bypassing and isolating the transfer equipment shall be permitted.

    a. True   b. False

16. Grounded circuit conductors for a legally required standby power source shall not be permitted to be connected to a grounding electrode conductor that is remotely located from the legally required standby power source.

    a. True   b. False

17. The authority having jurisdiction shall determine if an additional service for a legally required standby system shall be permitted.

    a. True   b. False

18. Type OFNR optical fiber cables that are no longer in use shall remain in place.

    a. True   b. False

19. Excluding transformers, the power source for Class 1 circuits shall have a maximum output of 2500 volt-amperes.

    a. True   b. False

20. Optional standby systems are intended to supply all on-site general power needs automatically for a period of 1 hour.

    a. True   b. False

21. The power output of the source for Class 1 signaling circuits must be limited.

    a. True   b. False

22. Optional standby system wiring shall be permitted in the same raceways as general wiring.

    a. True   b. False

23. Feeder circuit wiring for emergency systems shall be installed in areas that are completely protected by an approved automatic fire suppression system.

    a. True   b. False

24. Fiber optical cables shall be permitted in the same raceway with power-limited fire alarm systems conductors.

    a. True   b. False

25. When a portable optional standby power source is used as a non-separately derived system, the equipment grounding conductor shall be bonded to the system grounding electrode.

    a. True   b. False

26. An electric power source that operates in parallel with an electric primary power source and is capable of delivering energy to that source is known as a separately derived system.

    a. True   b. False

27. Conductors for circuits that operate at 50 volts or less and that supply two appliances shall not be smaller than #12 AWG copper.

    a. True   b. False

28. For Class 2 and Class 3 installations, cables installed in plena shall be Type CL2P or CL3Pplena.

    a. True   b. False

Special Conditions   189

29. An interconnected electric power production source shall operate in parallel with a primary power source of electricity.

   a. True   **b.** False

30. Sensors that indicate ground faults in a solidly grounded wye emergency system over 150 volts to ground shall be located at or ahead of the main system disconnecting means for the emergency source.

   a. True   **b.** False

31. All electric power production sources of an interconnected power production source must have a disconnection means.

   a. True   **b.** False

32. Without exception, if ground fault protection is used for an interconnect power production source, then the input of the system shall be connected to the output side of the ground fault protection.

   a. True   **b.** False

33. Class 2 and Class 3 circuits shall only be permitted in the same outlet boxes with Class 1 conductors if the conductors operate at less than 150 volts to ground.

   a. True   **b.** False

34. Other than power supplied from transformers, in a Class 1 circuit the product of the I_{max} and the V_{max} shall not exceed 1000 volt-amperes as determined with bypassing an overcurrent protection device.

   a. True   **b.** False

35. A 3-phase power production source must be able to be automatically disconnected from all ungrounded conductors of an interconnected power system if one of the phases of that source closes.

   a. True   **b.** False

36. Other than transformers, power sources that supply a Class 1 system shall be protected by overcurrent devices.

   a. True   **b.** False

37. On a wiring system that operates at 50 volts or less, only lampholders rated at 600 volts or less shall be used.

   a. True   **b.** False

38. Disregarding any exceptions, overcurrent protection for Class 1 conductors shall not exceed 7 amps for #18 AWG or 10 amps for #16 AWG.

    a. True   **b.** False

39. On a wiring system that operates at 50 volts or less, receptacles of not less than 15 amps shall be used.

    a. True   **b.** False

40. Wiring for two or more emergency circuits that are supplied from the same source shall not be permitted to run in the same cable box.

    a. True   **b.** False

41. A Class 1 circuit is the portion of a wiring system between the load side of an overcurrent device and the connected equipment that is rated at an output of not more than 30 volts and 600 amps.

    a. True   **b.** False

42. On a wiring system which operates at 50 volts or less, receptacles of not less than 15 amps shall be used.

    a. True   **b.** False

43. If a generator set requires a battery charger it shall be connected to the emergency system.

    a. True   **b.** False

44. Type CL2P cables and plenum signaling raceways shall be permitted for Class 2 circuits that are installed in spaces used for environmental air.

    a. True   **b.** False

45. If Class 3 conductors are labeled on the wire indicating a Class 3 rating, then circuits do not need to be labeled at terminals or junction locations.

    a. True   **b.** False

46. Room thermostats or water temperature regulating devices are two examples of controls that are considered approved safety control equipment for Class 1, 2, or 3 wiring systems.

    a. True   **b.** False

47. Only rigid metal conduit shall be used for Class 1, Class 2, and Class 3 wiring installations in order to protect remote-control circuits from physical damage.

    a. True   **b.** False

Special Conditions

48. Overcurrent devices used to protect Class 1 wiring installations shall be rated at not more than 167 percent of the volt-ampere rating of the source divided by the rated voltage.

    a. True   **b.** False

49. Electric power production system outputs for interconnected power production sources must be interconnected at the building service disconnecting means.

    a. True   **b.** False

50. When calculating Class 1 installation requirements I_{max} is the maximum output current load, including short circuit, and based on the overcurrent protection devices being bypassed.

    a. True   **b.** False

51. If a fuel cell system is used to power an emergency system, then it must be capable of supplying and maintaining the total load demand for a minimum of 1 1/2 hours.

    a. True   **b.** False

52. Current-limiting impedance shall not be bypassed when determining I_{max} for Class 1 installations.

    a. True   **b.** False

53. When calculating Class 1 installation requirements, V_{max} is the maximum output voltage regardless of the load and with rated input applied.

    a. True   **b.** False

54. Fire alarm cable that is not terminated at equipment other than a connector, or that is not identified with a "for future use" tag is considered to be abandoned.

    a. True   **b.** False

55. Class 1 remote-control circuits shall be greater than 600 volts.

    a. True   **b.** False

56. Audible or visual signals are not required for legally required standby systems.

    a. True   **b.** False

57. Overcurrent devices for Class 1 systems shall be located where the protected conductor receives its power supply.

    a. True   **b.** False

58. Fiber optical cables shall be permitted in the same cable tray with Class 2 and Class 3 power-limited circuits.

   a. True   **b.** False

59. Two or more Class 1 circuits shall be allowed to occupy the same cable tray regardless of whether the individual circuits are AC or DC, as long as all of the conductors are insulated for the maximum voltage of any given conductor in the cable tray.

   a. True   **b.** False

60. Type OFC optical fiber cable shall not be permitted to be installed in cable trays.

   a. True   **b.** False

61. Fiber optical cables shall be permitted in the same enclosure with Class 2 and Class 3 remote-control circuits.

   a. True   **b.** False

62. An optional standby system supplies power to public or private facilities where life and safety do not depend on the performance of the system.

   a. True   **b.** False

63. Class 1 circuit conductors that are 14 AWG or larger and are tapped from the load side of overcurrent protection devices of a controlled light and power circuit only require short-circuit and ground fault protection.

   a. True   **b.** False

64. Type OFN nonconductive and conductive fiber cables are approved for use in risers and plena.

   a. True   **b.** False

65. A generator for an interconnected power production source that operates in parallel with an electric supply system must have a compatible voltage, wave shape, and frequency with that supply system.

   a. True   **b.** False

66. Two or more Class 1 circuits shall be allowed to occupy the same cable tray only when the equipment being powered is functionally associated.

   a. True   **b.** False

67. Switches for emergency lighting circuits must be readily accessible.

   a. True   **b.** False

Special Conditions

68. In a closed-loop or programmed power distribution system, outlets shall be de-energized when a ground-fault condition exists.

    a. True   **b.** False

69. If a signaling circuit of 24 volts or less is protected by an overcurrent device, then current required shall not exceed 2 amperes.

    a. True   **b.** False

70. Fiber optical cables shall be permitted in the same enclosure with Class 2 and Class 3 remote-control circuits.

    a. True   **b.** False

71. Fiber optical cables shall be permitted in the same cable tray with community antenna television and radio systems conductors.

    a. True   **b.** False

72. Fire alarm circuits are classified as either non-power limited or power limited.

    a. True   **b.** False

73. Hybrid cabling of power communications and signaling conductors shall not be permitted under a common jacket.

    a. True   **b.** False

74. A fire alarm control conductor is that part of a wiring system located between the power side of an overcurrent device and connected equipment and the circuits that are powered and controlled by a fire alarm system.

    a. True   **b.** False

75. In a closed-loop or programmed power distribution system, outlets shall be de-energized when an overcurrent condition exists.

    a. True   **b.** False

76. General purpose optical fiber raceways shall be waterproof.

    a. True   **b.** False

77. When calculating Class 1 installation requirements, V_{max} is the maximum volt-ampere output achieved after one minute of operation, regardless of the load and after bypassing any current-limiting impedance.

    a. True   **b.** False

78. Individual emergency unit equipment shall include a battery that can maintain no less than 87 1/2 percent of the nominal battery voltage for the total lamp load of the unit for at least 1 1/2 hours.

    a. True   **b.** False

79. Conductors for circuits that operate at 50 volts or less shall not be smaller than #12 AWG copper.

    a. True   **b.** False

80. Type OFCR nonconductive and conductive fiber cables are approved for use in a vertical in shafts or from floor to floor.

    a. True   **b.** False

81. Type OFN cable shall not be installed in hazardous (classified) locations.

    a. True   **b.** False

82. Optical fiber cables with non-current carrying members shall be grounded.

    a. True   **b.** False

83. Class 1 circuit conductors that are 14 AWG or larger and are tapped from the load side of overcurrent protection devices of a controlled light and power circuit are permitted to be protected by the branch-circuit overcurrent protection devices rated no more than 200 percent of the ampacity of the Class 1 conductors.

    a. True   **b.** False

84. Type OFNP is not a fire resistive rated cable.

    a. True   **b.** False

85. A legally required standby alternate power source shall be permitted to also supply optional standby system loads as long as automatic selective load pickup and load shedding is provided to ensure there is adequate power to the legally required standby circuits.

    a. True   **b.** False

86. Composite optical fiber cables that only have current-carrying conductors for Class 1 circuits of 600 volts or less shall not occupy the same cabinet as conductors for electric light or power circuits operating at 600 volts or less.

    a. True   **b.** False

Special Conditions **195**

87. Emergency system current supplies shall provide power within 30 seconds of any failure of the normal supply to a building.

    a. True   **b.** False

88. PLFA is an abbreviation for Power Line Feeder Amperes.

    a. True   **b.** False

89. Fire alarm circuits shall be identified at terminals only.

    a. True   **b.** False

90. In a closed-loop or programmed power distribution system, outlets shall be de-energized when a nominal-operation acknowledgement signal is not received from the equipment connected to the outlet.

    a. True   **b.** False

91. Type OFNP cable shall be permitted to be installed in hazardous (classified) locations.

    a. True   **b.** False

92. Receptacles on a closed-loop power distribution system are interchangeable with other outlets.

    a. True   **b.** False

93. Wiring from an emergency source to emergency loads shall be permitted to run with wiring from a normal power source located in a transfer equipment enclosure.

    a. True   **b.** False

94. Non-conductive optical fiber cables shall not be allowed to occupy the same panel as non-power limed fire alarm circuits.

    a. True   **b.** False

95. Fiber optical cables shall be permitted in the same enclosure as Class 2 and Class 3 low-power network-powered broadband communications circuits.

    a. True   **b.** False

96. Class 1 circuits shall be permitted to be installed in manholes as underground conductors as long as they are in a metal-enclosed cable or Type UF cable.

    a. True   **b.** False

97. Optical fiber cables installed in risers with vertical runs in a shaft shall be permitted to be Type OFCR or OFN.

    a. True   **b.** False

98. Derating factors shall be applied only to Class 1 conductors that carry loads of less than 10 percent of the ampacity of each conductor and when there are four or more conductors in the power system.

    a. True   **b.** False

99. Class 2 and Class 3 circuits shall be permitted to be run in the same common raceway as Class 1 circuits.

    a. True   **b.** False

100. Class 2 and Class 3 cables installed in hazardous classified locations shall be Type PLTC.

    a. True   **b.** False

ANSWERS

QUIZ 1

1. **A** Section 760.21
2. **B** Section 770
3. **C** Section 725.(A)
4. **A** Section 701.11
5. **D** Section 760.21
6. **C** Section 760.3(A)
7. **B** Section 702
8. **A** Section 700.12
9. **A** Section 700.5
10. **C** Section 700.8(A)
11. **A** Section 700.15
12. **C** Sections 700.7(D) and 700.26
13. **B** Section 700.17
14. **C** Section 700.5
15. **C** Section 760.81(A)
16. **A** Section 760.81(G)
17. **B** Section 760.28(A)
18. **B** Section 760.56(C)
19. **A** Section 727.5
20. **B** Section 760.21

QUIZ 2

1.	B	26.	B	51.	B	76.	B
2.	B	27.	B	52.	A	77.	B
3.	A	28.	A	53.	A	78.	A
4.	A	29.	A	54.	A	79.	A
5.	A	30.	A	55.	B	80.	A
6.	B	31.	A	56.	B	81.	B
7.	B	32.	A	57.	A	82.	A
8.	A	33.	B	58.	A	83.	B
9.	B	34.	B	59.	A	84.	B
10.	A	35.	B	60.	B	85.	A
11.	B	36.	A	61.	A	86.	B
12.	A	37.	B	62.	A	87.	B
13.	B	38.	A	63.	A	88.	B
14.	B	39.	A	64.	B	89.	B
15.	A	40.	B	65.	A	90.	A
16.	B	41.	B	66.	A	91.	A
17.	A	42.	A	67.	B	92.	B
18.	B	43.	A	68.	A	93.	A
19.	A	44.	A	69.	B	94.	A
20.	B	45.	B	70.	A	95.	A
21.	B	46.	B	71.	A	96.	A
22.	A	47.	B	72.	A	97.	B
23.	A	48.	A	73.	B	98.	B
24.	A	49.	A	74.	B	99.	B
25.	A	50.	A	75.	A	100.	A

—NOTES—

Chapter 8
COMMUNICATIONS

Electrician's Exam Study Guide

There are only four articles in Chapter eight of the NEC® and they all deal with communication systems. Article 800 outlines communications circuit wiring, and Article 810 is all about radio and television equipment and antenna circuits in particular. Article 820 gets more specific and covers community antenna television and radio distribution systems and outlines coaxial cable circuits. Last, but in today's world certainly not least, Article 830 is on network-powered broadband communications systems. The focus is on systems that use power from the network.

Maybe at some point in the not-too-distant future, twisted pair telephone wiring will have gone the way of the party lines, rotary phones, and live operator assistance. Until then, there are still enough applications out there that use twisted pair wiring. Article 800 includes installation and protection means for the outside wiring of fire and burglar alarms, central station systems, hybrid power and communication cables, as well as telephone and telegraph systems. Section **[800.24]** instructs you to install communication components in a neat and workmanlike manner and to support exposed cables with straps or staples. Often you will find that telephone lines are supported by the same pole as electric lines, or that they all run parallel to each other. In theses cases, Section **[800.44(A)(2)]** prohibits communication lines from being attached to the same crossarms as power lines. If the communication lines are run underground through raceways, handholes or manholes where electric lines are present, then they must be separated from the power lines by brick, concrete, or some other suitable barrier as required by Section **[800.47(A)]**. The point at which communication lines enter a building must be protected by using an insulating bushing or a metal raceway. The specifications and exceptions for this entry point are listed in Section **[800.47(C)]**. If communication lines run overhead, they must be separated from power lines by 12 inches.

Communications wiring must be grounded. Section **[800.100(A)(3)]** states that the grounding conductor cannot be any smaller than 14 AWG. When working on a single-family or two-family dwelling unit, Section **[800.100(A)(4)]** indicates that the length of the primary grounding conductor has to be as short as possible and cannot exceed 20 feet.

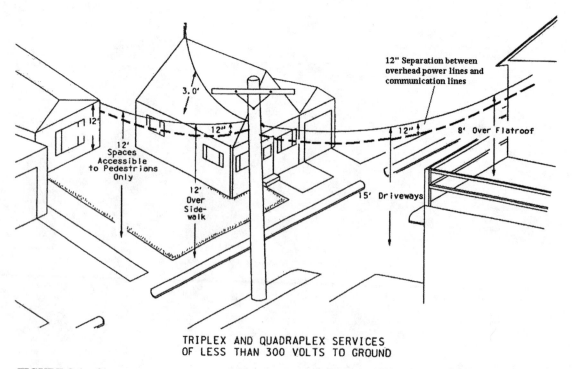

FIGURE 8.1 Clearance.

Communications

> **! Code update** SECTION 800.100(A)(4) FPN
>
> This FPN was added in 2005, explaining that a similar limitation of length as that listed in 800.100(A)(4) should apply to apartment and commercial buildings and will reduce undesirable voltage differences during lightning strikes.

This article refers you to the requirements of Article 250 for installation of required grounding electrodes. If a building doesn't have an existing grounding means, then Section **[800.100(B)(2)]** instructs you to drive a grounding rod that is at least 5 feet long and 1/2 inch in diameter into permanently damp earth. Regardless of the grounding means, communication grounding electrodes and power grounding electrode systems must be connected by a bonding jumper that cannot be smaller than 6 AWG, as outlined in Section **[800.100(C)]**. Primary protection grounding and bonding jumper requirements for mobile homes have their own installation rules, which are listed separately in Section **[800.100(A)(3)]**

Table **[800.113]** lists the required cable markings for various types of communications cables. Communications lines installed in a raceway have to meet the requirements for raceways detailed in NEC® Chapter three, though not spelled out in the NEC, they have to conform with the BICSI Cabling Installation Manual.

> **! Code update** SECTION 800.133(A)(1)(c) Exception 1
>
> This section in the 2005 Code was Section 800.52(A)(1)(c) Exception 1 in the 2002 edition. While communications wiring must still be kept separate from power wiring, that new wording in the exception says that power wiring and communications wiring can be in the same box if the two types of wiring are separated by a *permanent barrier or listed divider*.

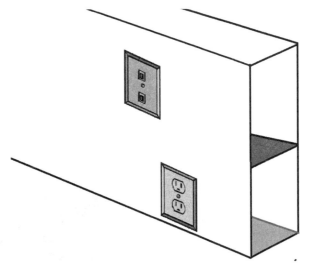

FIGURE 8.2 Communication and power wiring are seperated by a permanent barrier.

The types of approved cables and their applications are listed in Section **[800.154]**; however, some substitutions are allowed as illustrated in Table **[800.154]**. Communications wiring and cables cannot have a voltage rating over 300 volts and they must be copper unless you are running coax cable. Installation requirements are listed by cable type throughout Section **[800.179]**.

The next article, Article 810, covers radio, amateur radio, and television antennas, including vertical and dish antennas and the wiring that connects them to equipment. Metal antennas and masts need to be grounded as laid out in Section **[810.21]**; the required sizes for receiving station antenna conductors is listed in Table **[810.16(A)]**, and specific clearance requirements from conductors to receiving station buildings is listed in Section **[810.18]**. Grounding electrode conductors used for antennas must meet the standards listed in Section **[810.24]**, and approved connection methods found here include a metallic service raceway, service equipment enclosures, and grounded interior pipes that are within 5 feet of the grounding conductor entrance point to a building. Amateur radio transmission and receiving station antenna requirements begin with Section **[810.51]**, which, like the previous section, includes Table **[810.52]** for approved conductor sizes, Section **[810.55]** for entrance specifications, and Section **[810.58]** on grounding conductor installation requirements.

The next part of this chapter, Article 820, focuses on the distribution of television and radio signals within a facility through cable, rather than their transmission or reception via antennas. These signals are limited-energy, but they are high-frequency. Various clearance requirements, based on the type of attachment points, such as poles, masts and roof, are individually described in Section **[820.44]**. Underground coax that runs in the same manholes or ducts as power cabling has to be separated from the power lines by a suitable barrier, and Section **[820.47]** states direct buried coax must also be separated from power cabling by 12 inches, with two exceptions, which are listed. The outer conductive shield on coax cables has to be

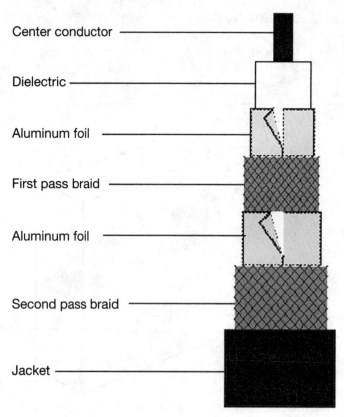

FIGURE 8.3 Coaxial cable.

Communications

grounded at the building as close as possible to the point where the cabling enters a building. Section **[820.93]** lays out the specifications for grounding coax to mobile home service equipment, and includes a requirement that the grounding location must be in sight from and not more than 30 feet away from a mobile home. The coax grounding conductor cannot be any smaller than 14 AWG and is not required to exceed 6 AWG. As was true for communications wiring in Section 800, the length of the primary coax grounding conductor has to be as short as possible and cannot exceed 20 feet. Grounding electrode system requirements for coax are found in Section **[820.100(B)]**, and a list of approved coax cable markings is provided in Table **[820.113]**. Just like Section 800, a separate segment on grounding and bonding for mobile homes is found in Section **[820.106]**

There is a provision for alternate wiring methods, which are later listed in Article 830.

Article 830 contains the installation requirements for network-powered broadband communications circuits that provide voice, audio, video, data, and interactive services through a network interface unit (NIU). Table **[830.15]** lists limitations for network-powered broadband communications systems based on the network power source. The various types of acceptable circuits for medium power and low-power broadband installations are provided in Sections **[830.40(A) & (B)]**. Minimum cover requirements for direct-buried network-powered installations, including those run in rigid metal conduit and nonmetallic raceways, are provided in Table **[830.47]**. Grounding conductors for network powered broadband systems must be at least 14 AWG, and grounding means are listed in Section **[830.100(B)(1)]**. A network powered broadband communications cable grounding terminal and the primary protection grounding terminal must be bonded together with a copper bonding conductor that is no less than 12 AWG, and allowances for mobile home grounding methods are listed in Section **[830.106(B)(1) & (2)]**. Both low and medium power network powered broadband circuit cables can be run in the same raceway, according to Section **[830.133(A)(1)]**, but neither can be run in the same raceway as power, electric light, NPLFA, or Class 1 conductors. The balance of Article 830 focuses on approved cabling types, applications and locations.

QUIZ 1

1. A circuit that extends from a communication utility service terminal and includes a NIU is considered to be which of the following:
   a. A network interface unit
   b. A block
   c. A network-powered broadband communications circuit
   d. A network power source

2. Communication components must be installed in a neat and workmanlike manner; consequently, the use of straps or staples on exposed cables is prohibited.
   a. True
   b. False

3. The bonding jumper for a communication grounding electrode and a power grounding electrode cannot be smaller than which of the following sizes:
   a. 6 AWG
   b. 8 AWG
   c. 14 AWG
   d. 16 AWG

4. If the communications lines are run underground through raceways, handholes or manholes where electric lines are present, then they must be separated from the power lines by which of the following:
   a. A suitable barrier
   b. A brick barrier
   c. A concrete barrier
   d. All of the above

5. If communications wiring is installed in a single-family home, which of the following requirements must be met:
   a. The wiring system must be grounded
   b. The grounding conductor cannot be smaller than 14 AWG
   c. The length of the primary grounding conductor cannot exceed 20 feet.
   d. All of the above

6. Low and medium power network powered broadband circuit cables cannot be run in the same raceway:
   a. True
   b. False

Communications

7. Communication grounding electrodes and power grounding electrode systems must be connected by which of the following:

    a. A grounded conductor   **b.** A junction box

    c. A bonding jumper   **d.** None of the above

8. In a single-family dwelling where CATVP cable is stapled alongside floor joists, the cable may be run through an air plenum.

    a. True   **b.** False

9. Communications wiring and cables cannot have a voltage rating over which of the following:

    a. 50 amps   **b.** 100 volts

    c. 120 volts   **d.** 300 volts

10. Grounding electrode conductors used for antennas may be connected to a metallic service raceway as a grounding means.

    a. True   **b.** False

11. Disregarding any exceptions, direct buried coax must also be separated from power cabling by which of the following distances:

    a. 6 inches   **b.** 12 inches

    c. 16 inches   **d.** 24 inches

12. Coax grounding conductors cannot be any smaller than which of the following sizes:

    a. 6 AWG   **b.** 8 AWG

    c. 12 AWG   **d.** 14 AWG

13. Grounding conductors for network powered broadband systems must be a minimum of which of the following sizes:

    a. 14 AWG   **b.** 16 AWG

    c. 18 AWG   **d.** None of the above

14. Wherever practicable, the minimum distance between overhead communications lines and lighting conductors should be which of the following:

    a. 3 feet   **b.** 5 feet

    c. 10 feet   **d.** None of the above

15. Bonding conductors for network powered broadband systems cannot be less than which of the following sizes:

    a. 6 AWG
    b. 8 AWG
    c. 12 AWG
    d. 14 AWG

16. The following types of communication wires shall be permitted to be installed in cable trays:

    a. CMP
    b. CMR
    c. CM
    d. All of the above

17. Where required by Code, coaxial grounding conductors shall be which of the following:

    a. Not smaller than #14 AWG wire
    b. Have a current-carrying capacity that is approximately the same as that of the outer conductor of the coaxial cable
    c. Not be required to exceed #6 AWG
    d. All of the above

18. Communication cabling for single family and two-family dwelling units shall be permitted to be which of the following:

    a. CMR
    b. CMP
    c. Both of the above
    d. None of the above

19. Type CMP communication cable may be substituted for which of the following communication cable types:

    a. CMR
    b. CMUC
    c. CSA
    d. Any of the above

20. Antennas and lead-in conductors shall be which of the following materials:

    a. Hand-drawn copper or bronze
    b. Aluminum alloy
    c. Copper-clad steel
    d. All of the above

21. Low power network-powered broadband communication system cabling shall be permitted in the same raceway or enclosure as which of the following:

    a. Power-limited fire alarm wiring
    b. Conductive optical fiber cables
    c. Medium power network-powered broadband communication system cabling
    d. All of the above

Communications

22. Indoor antennas shall not run closer than which of the following to other wiring system conductors on the premises:

   a. 2 inches

   b. 2 inches if the other conductors are located in cable armor

   c. 12 inches

   d. 3 inches if the other conductors are located in a metal raceway

23. Coaxial cable type CATVX with a diameter of less than 10 millimeters shall be permitted in which of the following installations:

   a. Single-family dwelling units

   b. Two-family dwellings

   c. Multi-family dwelling

   d. All of the above

24. Communication wires that are permitted in distributing frames and cross-connect arrays are which of the following:

   a. CMG and CMR

   b. CMP and CM

   c. Both of the above

   d. None of the above

25. Antenna discharge units inside a building shall be located in which of the following manners:

   a. Between the point of entry of the lead-in and the radio set

   b. As near as practicable to the entrance of the conductors to the building

   c. Not anywhere near a combustible material

   d. All of the above

26. A bonding jumper for receiving station installation shall be which of the following:

   I. Not smaller than 6 AWG copper

   II. Connected between the radio or television equipment grounding electrode and the structure grounding electrode

   III. Be connected between the equipment grounding electrode and power grounding electrode system at the building where separate electrodes are used

   IV. Not be required

   a. I only

   b. I and III

   c. IV only

   d. I and II

27. A cable positioned in such a manner that failure of supports or insulation could result in contact with another circuit is considered to be which of the following:
    a. Exposed
    b. Run in a straight line
    c. Accessible
    d. Unsupported

28. Communication cables installed under carpets shall be which of the following types:
    a. CMX
    b. CMUC
    c. Both of the above
    d. None of the above

29. Which of the following shall be permitted as a substitution for type BM network-powered broadband communication system cable:
    a. Type BMR
    b. Type BLR
    c. Type BLX
    d. All of the above

30. Bronze outdoor antenna conductors for receiving stations that must span 145 feet shall be which of the following AWG sizes:
    a. 12
    b. 14
    c. 17
    d. 19

31. Coaxial cable that is not in a raceway must have a separation of at least 4 inches for which of the above:
    a. Electric light and power wires
    b. Non-power limited fire alarm circuit conductors
    c. Class 1 cables
    d. All of the above

32. The clearance required between lead-in receiving station conductors and any non-bonded conductors that form part of a lightning rod system shall not be which of the following:
    a. Less than 6 feet
    b. Less than 12 inches
    c. More than 10 feet
    d. This type of connection is not permitted

33. Direct buried coaxial cable shall be separated from any light or power conductors or Class 1 circuits that are not installed in a raceway by which of the following means:
    a. By a tile or masonry barrier
    b. By at least 6 inches of clear space
    c. By at least 12 inches
    d. By use of a separate EMT or FMC conduit

Communications

34. Type CMX communication cabling is considered which of the following:
   a. General use
   b. Limited use
   c. Riser cable
   d. None of the above

35. If the outer conductive shield of coaxial cable is grounded, then which of the following shield grounding methods shall be used:
   a. A protective device that does not interrupt the grounding system shall be used
   b. A grounding conductor shall be used
   c. No additional protective devices are required
   d. None of the above

36. The grounding conductor for a receiving station shall not be smaller than which of the following:
   a. 17 AWG
   b. 10 AWG copper
   c. 8 AWG bronze
   d. None of the above

37. When coaxial cable is required by Code to have a grounding conductor, the conductor material shall be which of the following:
   a. Stranded or solid
   b. Copper
   c. Of a corrosion-resistant material
   d. All of the above

38. Communication cables run in raceways shall not have which of the following installation methods employed:
   a. Be strapped
   b. Be taped
   c. Be attached to the exterior of a raceway as a means of support
   d. All of the above

39. Which of the following network-powered broadband communication system cable types shall be permitted to run in ducts, plena and other spaces used for environmental air:
   a. Type BMR
   b. Type BLR
   c. Type BLP
   d. None of the above

40. Coaxial cable shall not be required to be listed and marked in which of the following cases:
    I. The length of cable run inside a building is 50 feet or less
    II. The cable is size #12 AWG or smaller
    III. The coax cable enters a building from the outside and is terminated at a grounding block.
    IV. The cable is type CATVP
    a. I only
    b. I and III
    c. II and IV
    b. All of the above

41. Which of the following communication cables are allowed in cable trays:
    a. CMG and CMR
    b. CMP and CM
    c. Both of the above
    d. None of the above

42. Which of the following has the greatest fire-resistance rating:
    a. CATVP
    b. CATV
    c. CATVR
    d. CATVX

43. Communication cables installed in nonconcealed spaces shall meet which of the following requirements:
    a. Be type CMX
    b. Have an exposed length of not more than 10 feet
    c. Both of the above
    d. None of the above

44. Which of the following coaxial cable types is permitted in vertical runs in a shaft:
    a. CATVP
    b. CATVR
    c. CATVX
    d. All of the above

45. The circuit that extends from a communication utility service terminal to a network interface unit is considered which of the following:
    a. The premises' network-powered wiring system
    b. The point of entrance
    c. The network-powered broadband communication circuit
    d. The fault protection device

Communications

46. The grounding conductor for a receiving station shall be connected to the nearest accessible location on which of the following:

   a. The building grounding electrode system

   b. A grounded interior metal water pipe within 10 feet of the point of entry of the building

   c. The interior location of the power service means

   d. All of the above

47. The grounding conductor for network-powered broadband communication system cabling shall be no larger than which of the following:

   a. 6 AWG
   b. 8 AWG
   c. 12 AWG
   d. 14 AWG

48. Communication cables installed in vertical runs and penetrate more than one floor shall be which of the following types:

   a. CMP
   b. CM
   c. CMR
   d. CMG

49. Types BMU, BM and BMR medium network-powered broadband communication system cables shall be factory assembled and shall comply with which of the following:

   a. A jacketed coaxial cable

   b. A jacketed combination of coaxial cable and multiple individual conductors

   c. Insulation on individual conductors shall be rated for a minimum of 300 volts

   d. All of the above

50. If the current supply is from a transformer, than coaxial cable is permitted to deliver low-energy to radio frequency distribution equipment operating at which of the following:

   a. 50 volts or less
   b. 60 volts or less
   c. 35 volts or less
   d. None of the above

QUIZ 2

1. Installed communications cable that is not terminated at one end and is tagged for future use is considered to be abandoned.
   - **a.** True
   - **b.** False

2. Network-powered broadband communication systems shall be classified as having low power sources.
   - **a.** True
   - **b.** False

3. If communication wires are run over a dwelling roof or other structure, such as a garage or storage shed, then there shall be at least 8 feet of clearance between the wires and the roof.
   - **a.** True
   - **b.** False

4. Communication cables that could accidentally come in contact with electric light and power conductors attached to buildings, and that operate at over 300 volts to ground shall be separated from woodwork by being supported on an insulating material such as glass or porcelain.
   - **a.** True
   - **b.** False

5. If a primary protector is installed inside a building where communication wires enter the building, then an insulating bushing is required.
   - **a.** True
   - **b.** False

6. An insulating bushing is not required where communication wires enter a building if the wires enter through concrete.
   - **a.** True
   - **b.** False

7. More than one communication wire shall not enter a building through any single raceway.
   - **a.** True
   - **b.** False

8. Type CATVP shall be permitted to be installed in raceways.
   - **a.** True
   - **b.** False

9. Interbuildings in geographical areas found along the Pacific coast that have an average of five or less thunderstorm days a year and earth resistivity of less than 85 ohm-meters are considered to not be exposed to lightning and shall not require a primary protector for communication cabling.
   - **a.** True
   - **b.** False

Communications

10. Network-powered broadband communication system cables shall be permitted to be direct buried under the concrete parking lot of a grocery store provided the concrete is a minimum of 4 inches thick and the cable is buried at least one foot deep:

   a. True   b. False

11. Fuseless primary protectors for communication cables are permitted under certain conditions.

   a. True   b. False

12. If fused-type primary protectors are used for communication cabling, they shall consist of a fuse in series with each line conductor.

   a. True   b. False

13. Secondary protectors installed on exposed communication circuits shall not be used without primary protectors.

   a. True   b. False

14. Type CATVP cable shall be permitted to be installed in nonconcealed spaces if the length of the exposed cable does not exceed 10 feet.

   a. True   b. False

15. At the point of entrance to a building, or as close thereto as is practicable, the metal sheath of communication cables shall be grounded or interrupted by an insulating joint or similar device.

   a. True   b. False

16. CMG and CMR communication wires are permitted in distributing frames and cross-connect arrays.

   a. True   b. False

17. Grounding conductors used for communication cable installations shall be of a corrosion-resistant material.

   a. True   b. False

18. Antenna discharge units shall be grounded.

   a. True   b. False

19. The primary protector grounding conductor for communication cabling in a one-family home shall not be longer than 20 feet.

   a. True   b. False

20. If a building does not have a standard grounding means available for communication cabling, then a metal structure or ground rod shall be permitted as a grounding means.

  a. True   **b.** False

21. CMP is an acceptable cable marking for communication wiring.

  a. True   **b.** False

22. Communication cables shall not be permitted in the same raceway as conductive fiber optic cables.

  a. True   **b.** False

23. Class 3 circuits that are run in the same cable with communications circuits shall be classified as communication circuits.

  a. True   **b.** False

24. Type CMX communication cable shall be permitted to be installed in a two-family dwelling as long as the cable diameter is less than 1/2 inch.

  a. True   **b.** False

25. Type CMP communication cable may be substituted for type CMG communication cable.

  a. True   **b.** False

26. Outdoor antenna conductors for receiving stations that are hand drawn copper and must span 100 feet shall be size 14 AWG.

  a. True   **b.** False

27. If an aerial wire is not confined within a block, then a listed primary protector must be provided on each communication circuit run.

  a. True   **b.** False

28. Antenna grounding conductors shall be insulated.

  a. True   **b.** False

29. Communication service drops shall be separated from service supply drops by a minimum of 12 inches at all points throughout the span.

  a. True   **b.** False

30. Communication cable raceways shall have an approved service head.

    a. True   **b.** False

31. A single grounding conductor shall not be allowed for both protective and operating purposes for a receiving station.

    a. True   **b.** False

32. A suitable barrier must separate communication cables from electric light and power lines or from Class 1 cables in a handhole.

    a. True   **b.** False

33. Without exception, coaxial cable shall be separated from lightning conductors by at least 6 feet.

    a. True   **b.** False

34. The grounding conductor for communication cabling systems shall not be smaller than 14 AWG.

    a. True   **b.** False

35. The communications grounding electrode and the power grounding electrode system at a building where separate electrodes are used shall have a bonding jumper not smaller than #8 AWG copper.

    a. True   **b.** False

36. Underground coaxial cables in a pedestal that also contains electric power conductors shall be separated from those conductors by a suitable barrier.

    a. True   **b.** False

37. Primary protectors for communication wiring shall be located as close as is practicable to the point of entrance of the wiring.

    a. True   **b.** False

38. Electrodes are bonded together to limit the potential differences between the electrodes and their associated wiring systems.

    a. True   **b.** False

39. Communication riser cable, marked as CMR, shall be permitted for communication cabling.

    a. True   **b.** False

40. The grounding conductor for communication cabling systems only needs to be listed for grounding purposes.

    a. True   **b.** False

41. Coaxial cables shall be permitted in the same enclosure as jacketed conductive optical fiber cables.

a. True   **b.** False

42. Type CMP and CM communication cables shall be permitted in cable trays.

a. True   **b.** False

43. Type CMX communication cable shall be permitted to be installed in nonconcealed spaces in a single family dwelling as long as the cable diameter is less than 0.25 inches.

a. True   **b.** False

44. When run on the outside of a raceway, coaxial cable must be supported by means of strapping, taping or other means to the exterior of the raceway.

a. True   **b.** False

45. A typical basic network-powered broadband communication system includes a cable that supplies power and broadband signal to a network interface unit, which converts the power to component signals.

a. True   **b.** False

46. Each interbuilding communication circuit on premises that have lightning exposure shall be protected by a listed primary protector at the load end of the interbuilding circuit.

a. True   **b.** False

47. The grounding conductor for communication wiring shall not be permitted to be connected to a metallic power service raceway.

a. True   **b.** False

48. Communication cables shall be permitted in the same raceway as Class 2 power-limited cables.

a. True   **b.** False

49. Secondary protectors for indoor communication wires must be capable of safely limiting currents to no less than the current-carrying capacity of the communication cabling.

a. True   **b.** False

50. Network-powered broadband communication circuits located at the service side of a network interface unit are defined as the "premise wiring"

a. True   **b.** False

ANSWERS

QUIZ 1

1. **C** Section 830.2
2. **B** Section 800.24
3. **A** Section 800.100(C)
4. **D** Section 800.47(A)
5. **D** Section 800.100(A)(3) and (4)
6. **B** Section 830.133(A)(1)
7. **C** Section 800.100(C)
8. **A** Section 820.154(A)
9. **D** Section 800.179
10. **A** Section 810.24
11. **B** Section 820.47
12. **D** Section 820.93
13. **A** Section 830.100(A)(3)
14. **D** Section 800.53
15. **C** Section 830.106(B)(1)
16. **D** Section 800.133(2)(B)
17. **D** Section 820.100(A)(3)
18. **D** Section 800.15 (B)(3)
19. **A** Table 800.154
20. **D** Section 810.11
21. **C** Section 830.133 A)(1)(b)
22. **B** Section 810.18(B)
23. **D** Section 820.154(D)(4) and (5)
24. **C** Section 800.154(C)
25. **D** Section 810.20(B)
26. **B** Section 810.21(J)
27. **A** Section 820.2 Definition of Exposed
28. **B** Section 800.154(E)(6)
29. **A** Table 830.133
30. **C** Table 800.16(A)
31. **D** Section 820.44(F)(1)
32. **A** Section 810.18(A)
33. **C** Section(B)
34. **B** Section 800.179(E)
35. **C** Section 820.93(A)
36. **B** Section 810.21(H)
37. **D** Section 820.100(A)(2)
38. **D** Section 800.133(2)(C)
39. **C** Section 830.154(B)
40. **B** Section 820.113 Exception 2
41. **C** Section 800.154(D)
42. **A** Table 820.113 FPN 1
43. **C** Section 800.154(E)(3)
44. **B** Section 820.154(B)(1)
45. **C** Section 830.2 Definitions
46. **A** Section 810.2(a)(F)(a)
47. **A** Section 830.100(3)
48. **C** Section 800.15 (B)(1)
49. **D** Section 830.179(A)(1)
50. **B** Section 820.15

QUIZ 2

1. B
2. B
3. B
4. A
5. A
6. B
7. B
8. B
9. B
10. B
11. A
12. B
13. A
14. B
15. A
16. A
17. A
18. A
19. A
20. A
21. A
22. B
23. A
24. B
25. A
26. A
27. A
28. B
29. A
30. A
31. B
32. A
33. B (NOTE: the code says "where practicable.")
34. A
35. B
36. A
37. A
38. A
39. A
40. B
41. A
42. A
43. B
44. B
45. B (NOTE: it converts the broadband signal to component signals.)
46. B
47. B
48. A
49. A
50. B

Chapter 9
TABLES, ANNEXES, AND EXAMPLES

In some ways, Chapter nine of the NEC® is like a quick reference library. Check out the tables in the beginning of the chapter and you'll find listings of the radius of conduits and tubing bends, the dimensions and percentage of area for various types of conduit, approved dimensions of insulated conductors and fixture wire, and even conductor size requirements based on direct current resistance.

One important thing to remember about the annexes is that they are not part of the code requirements; rather they are included for your information and reference. For example, Table 8, Conductor Properties, which is shown below, provides a comprehensive quick reference for conductor sizes, diameters, and types.

Table 8 Conductor Properties

									Direct-Current Resistance at 75°C (167°F)					
		Conductors							Copper					
		Stranding			Overall				Uncoated		Coated		Aluminum	
Size (AWG or kcmil)	Area		Diameter		Diameter		Area							
	mm²	Circular mils	Quantity	mm	in.	mm	in.	mm²	in.²	ohm/km	ohm/kFT	ohm/km	ohm/kFT	ohm/km	ohm/kFT
18	0.823	1620	1	—	—	1.02	0.040	0.823	0.001	25.5	7.77	26.5	8.08	42.0	12.8
18	0.823	1620	7	0.39	0.015	1.16	0.046	1.06	0.002	26.1	7.95	27.7	8.45	42.8	13.1
16	1.31	2580	1	—	—	1.29	0.051	1.31	0.002	16.0	4.89	16.7	5.08	26.4	8.05
16	1.31	2580	7	0.49	0.019	1.46	0.058	1.68	0.003	16.4	4.99	17.3	5.29	26.9	8.21
14	2.08	4110	1	—	—	1.63	0.064	2.08	0.003	10.1	3.07	10.4	3.19	16.6	5.06
14	2.08	4110	7	0.62	0.024	1.85	0.073	2.68	0.004	10.3	3.14	10.7	3.26	16.9	5.17
12	3.31	6530	1	—	—	2.05	0.081	3.31	0.005	6.34	1.93	6.57	2.01	10.45	3.18
12	3.31	6530	7	0.78	0.030	2.32	0.092	4.25	0.006	6.50	1.98	6.73	2.05	10.69	3.25
10	5.261	10380	1	—	—	2.588	0.102	5.26	0.008	3.984	1.21	4.148	1.26	6.561	2.00
10	5.261	10380	7	0.98	0.038	2.95	0.116	6.76	0.011	4.070	1.24	4.226	1.29	6.679	2.04
8	8.367	16510	1	—	—	3.264	0.128	8.37	0.013	2.506	0.764	2.579	0.786	4.125	1.26
8	8.367	16510	7	1.23	0.049	3.71	0.146	10.76	0.017	2.551	0.778	2.653	0.809	4.204	1.28
6	13.30	26240	7	1.56	0.061	4.67	0.184	17.09	0.027	1.608	0.491	1.671	0.510	2.652	0.808
4	21.15	41740	7	1.96	0.077	5.89	0.232	27.19	0.042	1.010	0.308	1.053	0.321	1.666	0.508
3	26.67	52620	7	2.20	0.087	6.60	0.260	34.28	0.053	0.802	0.245	0.833	0.254	1.320	0.403
2	33.62	66360	7	2.47	0.097	7.42	0.292	43.23	0.067	0.634	0.194	0.661	0.201	1.045	0.319
1	42.41	83690	19	1.69	0.066	8.43	0.332	55.80	0.087	0.505	0.154	0.524	0.160	0.829	0.253
1/0	53.49	105600	19	1.89	0.074	9.45	0.372	70.41	0.109	0.399	0.122	0.415	0.127	0.660	0.201
2/0	67.43	133100	19	2.13	0.084	10.62	0.418	88.74	0.137	0.3170	0.0967	0.329	0.101	0.523	0.159
3/0	85.01	167800	19	2.39	0.094	11.94	0.470	111.9	0.173	0.2512	0.0766	0.2610	0.0797	0.413	0.126
4/0	107.2	211600	19	2.68	0.106	13.41	0.528	141.1	0.219	0.1996	0.0608	0.2050	0.0626	0.328	0.100
250	127	—	37	2.09	0.082	14.61	0.575	168	0.260	0.1687	0.0515	0.1753	0.0535	0.2778	0.0847
300	152	—	37	2.29	0.090	16.00	0.630	201	0.312	0.1409	0.0429	0.1463	0.0446	0.2318	0.0707
350	177	—	37	2.47	0.097	17.30	0.681	235	0.364	0.1205	0.0367	0.1252	0.0382	0.1984	0.0605
400	203	—	37	2.64	0.104	18.49	0.728	268	0.416	0.1053	0.0321	0.1084	0.0331	0.1737	0.0529
500	253	—	37	2.95	0.116	20.65	0.813	336	0.519	0.0845	0.0258	0.0869	0.0265	0.1391	0.0424
600	304	—	61	2.52	0.099	22.68	0.893	404	0.626	0.0704	0.0214	0.0732	0.0223	0.1159	0.0353
700	355	—	61	2.72	0.107	24.49	0.964	471	0.730	0.0603	0.0184	0.0622	0.0189	0.0994	0.0303
750	380	—	61	2.82	0.111	25.35	0.998	505	0.782	0.0563	0.0171	0.0579	0.0176	0.0927	0.0282
800	405	—	61	2.91	0.114	26.16	1.030	538	0.834	0.0528	0.0161	0.0544	0.0166	0.0868	0.0265
900	456	—	61	3.09	0.122	27.79	1.094	606	0.940	0.0470	0.0143	0.0481	0.0147	0.0770	0.0235
1000	507	—	61	3.25	0.128	29.26	1.152	673	1.042	0.0423	0.0129	0.0434	0.0132	0.0695	0.0212
1250	633	—	91	2.98	0.117	32.74	1.289	842	1.305	0.0338	0.0103	0.0347	0.0106	0.0554	0.0169
1500	760	—	91	3.26	0.128	35.86	1.412	1011	1.566	0.02814	0.00858	0.02814	0.00883	0.0464	0.0141
1750	887	—	127	2.98	0.117	38.76	1.526	1180	1.829	0.02410	0.00735	0.02410	0.00756	0.0397	0.0121
2000	1013	—	127	3.19	0.126	41.45	1.632	1349	2.092	0.02109	0.00643	0.02109	0.00662	0.0348	0.0106

Notes:
1. These resistance values are valid **only** for the parameters as given. Using conductors having coated strands, different stranding type, and, especially, other temperatures changes the resistance.
2. Formula for temperature change: $R_2 = R_1 [1 + \alpha (T_2 - 75)]$ where $\alpha_{cu} = 0.00323$, $\alpha_{AL} = 0.00330$ at 75°C.
3. Conductors with compact and compressed stranding have about 9 percent and 3 percent, respectively, smaller bare conductor diameters than those shown. See Table 5A for actual compact cable dimensions.
4. The IACS conductivities used: bare copper = 100%, aluminum = 61%.
5. Class B stranding is listed as well as solid for some sizes. Its overall diameter and area is that of its circumscribing circle.

FPN: The construction information is per NEMA WC8-1992 or ANSI/UL 1581-1998. The resistance is calculated per National Bureau of Standards Handbook 100, dated 1966, and Handbook 109, dated 1972.

Tables, Annexes, and Examples

Additionally, Table 9 is a detailed listing of AC resistance and reactance for 600 volt cables that are 3-phase operating at 60 Hz, and where there are 3 single conductors in a conduit. As you can see in the table below, wire size and type examples are provided.

Table 9 Alternating-Current Resistance and Reactance for 600-Volt Cables, 3-Phase, 60 Hz, 75°C (167°F) — Three Single Conductors in Conduit

Ohms to Neutral per Kilometer
Ohms to Neutral per 1000 Feet

Size (AWG or kcmil)	X_L (Reactance) for All Wires		Alternating-Current Resistance for Uncoated Copper Wires			Alternating-Current Resistance for Aluminum Wires			Effective Z at 0.85 PF for Uncoated Copper Wires			Effective Z at 0.85 PF for Aluminum Wires			Size (AWG or kcmil)
	PVC, Aluminum Conduits	Steel Conduit	PVC Conduit	Aluminum Conduit	Steel Conduit	PVC Conduit	Aluminum Conduit	Steel Conduit	PVC Conduit	Aluminum Conduit	Steel Conduit	PVC Conduit	Aluminum Conduit	Steel Conduit	
14	0.190 / 0.058	0.240 / 0.073	10.2 / 3.1	10.2 / 3.1	10.2 / 3.1	—	—	—	8.9 / 2.7	8.9 / 2.7	8.9 / 2.7	—	—	—	14
12	0.177 / 0.054	0.223 / 0.068	6.6 / 2.0	6.6 / 2.0	6.6 / 2.0	10.5 / 3.2	10.5 / 3.2	10.5 / 3.2	5.6 / 1.7	5.6 / 1.7	5.6 / 1.7	9.2 / 2.8	9.2 / 2.8	9.2 / 2.8	12
10	0.164 / 0.050	0.207 / 0.063	3.9 / 1.2	3.9 / 1.2	3.9 / 1.2	6.6 / 2.0	6.6 / 2.0	6.6 / 2.0	3.6 / 1.1	3.6 / 1.1	3.6 / 1.1	5.9 / 1.8	5.9 / 1.8	5.9 / 1.8	10
8	0.171 / 0.052	0.213 / 0.065	2.56 / 0.78	2.56 / 0.78	2.56 / 0.78	4.3 / 1.3	4.3 / 1.3	4.3 / 1.3	2.26 / 0.69	2.26 / 0.69	2.30 / 0.70	3.6 / 1.1	3.6 / 1.1	3.6 / 1.1	8
6	0.167 / 0.051	0.210 / 0.064	1.61 / 0.49	1.61 / 0.49	1.61 / 0.49	2.66 / 0.81	2.66 / 0.81	2.66 / 0.81	1.44 / 0.44	1.48 / 0.45	1.48 / 0.45	2.33 / 0.71	2.36 / 0.72	2.36 / 0.72	6
4	0.157 / 0.048	0.197 / 0.060	1.02 / 0.31	1.02 / 0.31	1.02 / 0.31	1.67 / 0.51	1.67 / 0.51	1.67 / 0.51	0.95 / 0.29	0.95 / 0.29	0.98 / 0.30	1.51 / 0.46	1.51 / 0.46	1.51 / 0.46	4
3	0.154 / 0.047	0.194 / 0.059	0.82 / 0.25	0.82 / 0.25	0.82 / 0.25	1.31 / 0.40	1.35 / 0.41	1.31 / 0.40	0.75 / 0.23	0.79 / 0.24	0.79 / 0.24	1.21 / 0.37	1.21 / 0.37	1.21 / 0.37	3
2	0.148 / 0.045	0.187 / 0.057	0.62 / 0.19	0.66 / 0.20	0.66 / 0.20	1.05 / 0.32	1.05 / 0.32	1.05 / 0.32	0.62 / 0.19	0.62 / 0.19	0.66 / 0.20	0.98 / 0.30	0.98 / 0.30	0.98 / 0.30	2
1	0.151 / 0.046	0.187 / 0.057	0.49 / 0.15	0.52 / 0.16	0.52 / 0.16	0.82 / 0.25	0.85 / 0.26	0.82 / 0.25	0.52 / 0.16	0.52 / 0.16	0.52 / 0.16	0.79 / 0.24	0.79 / 0.24	0.82 / 0.25	1
1/0	0.144 / 0.044	0.180 / 0.055	0.39 / 0.12	0.43 / 0.13	0.39 / 0.12	0.66 / 0.20	0.69 / 0.21	0.66 / 0.20	0.43 / 0.13	0.43 / 0.13	0.43 / 0.13	0.62 / 0.19	0.66 / 0.20	0.66 / 0.20	1/0
2/0	0.141 / 0.043	0.177 / 0.054	0.33 / 0.10	0.33 / 0.10	0.33 / 0.10	0.52 / 0.16	0.52 / 0.16	0.52 / 0.16	0.36 / 0.11	0.36 / 0.11	0.36 / 0.11	0.52 / 0.16	0.52 / 0.16	0.52 / 0.16	2/0
3/0	0.138 / 0.042	0.171 / 0.052	0.253 / 0.077	0.269 / 0.082	0.259 / 0.079	0.43 / 0.13	0.43 / 0.13	0.43 / 0.13	0.289 / 0.088	0.302 / 0.092	0.308 / 0.094	0.43 / 0.13	0.43 / 0.13	0.46 / 0.14	3/0
4/0	0.135 / 0.041	0.167 / 0.051	0.203 / 0.062	0.220 / 0.067	0.207 / 0.063	0.33 / 0.10	0.36 / 0.11	0.33 / 0.10	0.243 / 0.074	0.256 / 0.078	0.262 / 0.080	0.36 / 0.11	0.36 / 0.11	0.36 / 0.11	4/0
250	0.135 / 0.041	0.171 / 0.052	0.171 / 0.052	0.187 / 0.057	0.177 / 0.054	0.279 / 0.085	0.295 / 0.090	0.282 / 0.086	0.217 / 0.066	0.230 / 0.070	0.240 / 0.073	0.308 / 0.094	0.322 / 0.098	0.33 / 0.10	250
300	0.135 / 0.041	0.167 / 0.051	0.144 / 0.044	0.161 / 0.049	0.148 / 0.045	0.233 / 0.071	0.249 / 0.076	0.236 / 0.072	0.194 / 0.059	0.207 / 0.063	0.213 / 0.065	0.269 / 0.082	0.282 / 0.086	0.289 / 0.088	300
350	0.131 / 0.040	0.164 / 0.050	0.125 / 0.038	0.141 / 0.043	0.128 / 0.039	0.200 / 0.061	0.217 / 0.066	0.207 / 0.063	0.174 / 0.053	0.190 / 0.058	0.197 / 0.060	0.240 / 0.073	0.253 / 0.077	0.262 / 0.080	350
400	0.131 / 0.040	0.161 / 0.049	0.108 / 0.033	0.125 / 0.038	0.115 / 0.035	0.177 / 0.054	0.194 / 0.059	0.180 / 0.055	0.161 / 0.049	0.174 / 0.053	0.184 / 0.056	0.217 / 0.066	0.233 / 0.071	0.240 / 0.073	400

Reprinted with permission from NFPA 70, *National Electrical Code*®, Copyright ©2004, National Fire Protection Association, Quincy, MA 02169. This reprinted material is not the complete and official position of the National Fire Protection Association on the referenced subject which is represented only by the standard in its entirety.

Table 12(A) PLFA Alternating-Current Power Source Limitations

Power Source		Inherently Limited Power Source (Overcurrent Protection Not Required)			Not Inherently Limited Power Source (Overcurrent Protection Required)		
Circuit voltage V_{max} (volts) (see Note 1)		0 through 20	Over 20 and through 30	Over 30 and through 100	0 through 20	Over 20 and through 100	Over 100 and through 150
Power limitations VA_{max} (volt-amperes) (see Note 1)		—	—	—	250 (see Note 2)	250	N.A.
Current limitations I_{max} (amperes) (see Note 1)		8.0	8.0	$150/V_{max}$	$1000/V_{max}$	$1000/V_{max}$	1.0
Maximum overcurrent protection (amperes)		—	—	—	5.0	$100/V_{max}$	1.0
Power source maximum nameplate ratings	VA (volt-amperes)	$5.0 \times V_{max}$	100	100	$5.0 \times V_{max}$	100	100
	Current (amperes)	5.0	$100/V_{max}$	$100/V_{max}$	5.0	$100/V_{max}$	$100/V_{max}$

Table 12(B) PLFA Direct-Current Power Source Limitations

Power Source		Inherently Limited Power Source (Overcurrent Protection Not Required)			Not Inherently Limited Power Source (Overcurrent Protection Required)		
Circuit voltage V_{max} (volts) (see Note 1)		0 through 20	Over 20 and through 30	Over 30 and through 100	0 through 20	Over 20 and through 100	Over 100 and through 150
Power limitations VA_{max} (volt-amperes) (see Note 1)		—	—	—	250 (see Note 2)	250	N.A.
Current limitations I_{max} (amperes) (see Note 1)		8.0	8.0	$150/V_{max}$	$1000/V_{max}$	$1000/V_{max}$	1.0
Maximum overcurrent protection (amperes)		—	—	—	5.0	$100/V_{max}$	1.0
Power source maximum nameplate ratings	VA (volt-amperes)	$5.0 \times V_{max}$	100	100	$5.0 \times V_{max}$	100	100
	Current (amperes)	5.0	$100/V_{max}$	$100/V_{max}$	5.0	$100/V_{max}$	$100/V_{max}$

Notes for Tables 12(A) and 12(B)

1. V_{max}, I_{max}, and VA_{max} are determined as follows:

V_{max}: Maximum output voltage regardless of load with rated input applied.

I_{max}: Maximum output current under any noncapacitive load, including short circuit, and with overcurrent protection bypassed if used. Where a transformer limits the output current, I_{max} limits apply after 1 minute of operation. Where a current-limiting impedance, listed for the purpose, is used in combination with a nonpower-limited transformer or a stored energy source, e.g., storage battery, to limit the output current, I_{max} limits apply after 5 seconds.

VA_{max}: Maximum volt-ampere output after 1 minute of operation regardless of load and overcurrent protection bypassed if used. Current limiting impedance shall not be bypassed when determining I_{max} and VA_{max}.

2. If the power source is a transformer, VA_{max} is 350 or less when V_{max} is 15 or less.

Reprinted with permission from NFPA 70, *National Electrical Code*®, Copyright ©2004, National Fire Protection Association, Quincy, MA 02169. This reprinted material is not the complete and official position of the National Fire Protection Association on the referenced subject which is represented only by the standard in its entirety.

Tables 12(A) and 12(B) show the required power source limitations for power-limited fire alarm systems. These fire alarm circuits must be either inherently limited, which requires non overcurrent protection, or not inherently limited, which requires a combination of power source and overcurrent protection.

Annex A provides a listing of product safety standards, while Annex B explains how to calculate and apply ampacity based on the number of conductors, the size of the wire, and the voltage of the system. Section B.**[310.15(B)(7)]** listed "Examples Showing Use of Figure B.310.1 for Electrical Duct Bank Ampacity Modifications." The explanation states the following:

> "Figure B.310.1 is used for interpolation or extrapolation for values of Rho and load factor for cables installed in electrical ducts. The upper family of curves shows the variation in ampacity and Rho at unity load factor in terms of I_1, the ampacity for Rho = 60 and 50 percent

load factor. Each curve is designated for a particular ratio I_2/I_1, where I_2 is the ampacity at Rho = 120 and 100 percent load factor.

The lower family of curves shows the relationship between Rho and load factor that will give substantially the same ampacity as the indicated value of Rho at 100 percent load factor.

As an example, to find the ampacity of a 500 kcmil copper cable circuit for six electrical ducts as shown in Table B.310.5: At the Rho = 60, LF = 50, I_1 = 583; for Rho = 120 and LF = 100, I_2 = 400. The ratio I_2/I_1, = 0.686. Locate Rho = 90 at the bottom of the chart and follow the 90 Rho line to the intersection with 100 percent load factor where the equivalent Rho = 90. Then follow the 90 Rho line to I_2/I_1, ratio of 0.686 where F = 0.74. The desired ampacity = 0.74 × 583 = 431, which agrees with the table for Rho = 90, LF = 100.

To determine the ampacity for the same circuit where Rho = 80 and LF = 75, using Figure B.310.1, the equivalent Rho = 43, F = 0.855, and the desired ampacity = 0.855 × 583 = 498 amperes. Values for using Figure B.310.1 are found in the electrical duct bank ampacity tables of the annex.

Where the load factor is less than 100 percent and can be verified by measurement or calculation, the ampacity of electrical duct bank installations can be modified as shown. Different values of Rho can be accommodated in the same manner."

Table B.310.1 Ampacities of Two or Three Insulated Conductors, Rated 0 Through 2000 Volts, Within an Overall Covering (Multiconductor Cable), in Raceway in Free Air Based on Ambient Air Temperature of 30°C (86°F)

	Temperature Rating of Conductor. (See Table 310.13.)						
	60°C (140°F)	75°C (167°F)	90°C (194°F)	60°C (140°F)	75°C (167°F)	90°C (194°F)	
	Types TW, UF	Types RHW, THHW, THW, THWN, XHHW, ZW	Types THHN, THHW, THW-2, THWN-2, RHH, RWH-2, USE-2, XHHW, XHHW-2, ZW-2	Type TW	Types RHW, THHW, THW, THWN, XHHW	Types THHN, THHW, THW-2, THWN-2, RHH, RWH-2, USE-2, XHHW, XHHW-2, ZW-2	
Size (AWG or kcmil)	COPPER			ALUMINUM OR COPPER-CLAD ALUMINUM			Size (AWG or kcmil)
14	16*	18*	21*	—	—	—	14
12	20*	24*	27*	16*	18*	21*	12
10	27*	33*	36*	21*	25*	28*	10
8	36	43	48	28	33	37	8
6	48	58	65	38	45	51	6
4	66	79	89	51	61	69	4
3	76	90	102	59	70	79	3
2	88	105	119	69	83	93	2
1	102	121	137	80	95	106	1
1/0	121	145	163	94	113	127	1/0
2/0	138	166	186	108	129	146	2/0
3/0	158	189	214	124	147	167	3/0
4/0	187	223	253	147	176	197	4/0
250	205	245	276	160	192	217	250
300	234	281	317	185	221	250	300
350	255	305	345	202	242	273	350
400	274	328	371	218	261	295	400
500	315	378	427	254	303	342	500

Reprinted with permission from NFPA 70, *National Electrical Code*®, Copyright ©2004, National Fire Protection Association, Quincy, MA 02169. This reprinted material is not the complete and official position of the National Fire Protection Association on the referenced subject which is represented only by the standard in its entirety.

Table B.**[310.7]** lists various ampacities, by wire size, of 3 single insulated conductors that are rated up to 2000 volts, and are installed in underground electrical ducts with 3 conductors per duct, and are based on an ambient temperature of 68 degrees F. Correction factors for other temperatures are included in the table.

Table B.310.7 Ampacities of Three Single Insulated Conductors, Rated 0 Through 2000 Volts, in Underground Electrical Ducts (Three Conductors per Electrical Duct) Based on Ambient Earth Temperature of 20°C (68°F), Electrical Duct Arrangement per Figure B.310.2, Conductor Temperature 75°C (167°F)

Size (AWG or kcmil)	1 Electrical Duct (Fig. B.310.2, Detail 1) Types RHW, THHW, THW, THWN, XHHW, USE COPPER			3 Electrical Ducts (Fig. B.310.2, Detail 2) Types RHW, THHW, THW, THWN, XHHW, USE COPPER			6 Electrical Ducts (Fig. B.310.2, Detail 3) Types RHW, THHW, THW, THWN, XHHW, USE COPPER			1 Electrical Duct (Fig. B.310.2, Detail 1) Types RHW, THHW, THW, THWN, XHHW, USE ALUMINUM OR COPPER-CLAD ALUMINUM			3 Electrical Ducts (Fig. B.310.2, Detail 2) Types RHW, THHW, THW, THWN, XHHW, USE ALUMINUM OR COPPER-CLAD ALUMINUM			6 Electrical Ducts (Fig. B.310.2, Detail 3) Types RHW, THHW, THW, THWN, XHHW, USE ALUMINUM OR COPPER-CLAD ALUMINUM			Size (AWG or kcmil)
	RHO 60 LF 50	RHO 90 LF 100	RHO 120 LF 100	RHO 60 LF 50	RHO 90 LF 100	RHO 120 LF 100	RHO 60 LF 50	RHO 90 LF 100	RHO 120 LF 100	RHO 60 LF 50	RHO 90 LF 100	RHO 120 LF 100	RHO 60 LF 50	RHO 90 LF 100	RHO 120 LF 100	RHO 60 LF 50	RHO 90 LF 100	RHO 120 LF 100	
8	63	58	57	61	51	49	57	44	41	49	45	44	47	40	38	45	34	32	8
6	84	77	75	80	67	63	75	56	53	66	60	58	63	52	49	59	44	41	6
4	111	100	98	105	86	81	98	73	67	86	78	76	79	67	63	77	57	52	4
3	129	116	113	122	99	94	113	83	77	101	91	89	83	77	73	84	65	60	3
2	147	132	128	139	112	106	129	93	86	115	103	100	108	87	82	101	73	67	2
1	171	153	148	161	128	121	149	106	98	133	119	115	126	100	94	116	83	77	1
1/0	197	175	169	185	146	137	170	121	111	153	136	132	144	114	107	133	94	87	1/0
2/0	226	200	193	212	166	156	194	136	126	176	156	151	165	130	121	151	106	98	2/0
3/0	260	228	220	243	189	177	222	154	142	203	178	172	189	147	138	173	121	111	3/0
4/0	301	263	253	280	215	201	255	175	161	235	205	198	219	168	157	199	137	126	4/0
250	334	290	279	310	236	220	281	192	176	261	227	218	242	185	172	220	150	137	250
300	373	321	308	344	260	242	310	210	192	293	252	242	272	204	190	245	165	151	300
350	409	351	337	377	283	264	340	228	209	321	276	265	296	222	207	266	179	164	350
400	442	376	361	394	302	280	368	243	223	349	297	284	321	238	220	288	191	174	400
500	503	427	409	460	341	316	412	273	249	397	338	323	364	270	250	326	216	197	500
600	552	468	447	511	371	343	457	296	270	446	373	356	408	296	274	365	236	215	600
700	602	509	486	553	402	371	492	319	291	488	408	389	443	321	297	394	255	232	700
750	632	529	505	574	417	385	509	330	301	508	425	405	461	334	309	409	265	241	750
800	654	544	520	597	428	395	527	338	308	530	439	418	481	344	318	427	273	247	800
900	692	575	549	628	450	415	554	355	323	563	466	444	510	365	337	450	288	261	900
1000	730	605	576	659	472	435	581	372	338	597	494	471	538	385	355	475	304	276	1000

Ambient Temp. (°C)	Correction Factors																		Ambient Temp. (°F)
6–10	1.09			1.09			1.09			1.09			1.09			1.09			43–50
11–15	1.04			1.04			1.04			1.04			1.04			1.04			52–59
16–20	1.00			1.00			1.00			1.00			1.00			1.00			61–68
21–25	0.95			0.95			0.95			0.95			0.95			0.95			70–77
26–30	0.90			0.90			0.90			0.90			0.90			0.90			79–86

Reprinted with permission from NFPA 70, *National Electrical Code*®, Copyright ©2004, National Fire Protection Association, Quincy, MA 02169. This reprinted material is not the complete and official position of the National Fire Protection Association on the referenced subject which is represented only by the standard in its entirety.

Tables, Annexes, and Examples

The next several tables include the same type of information, based on varying conditions. For example, Table B.**[310.10]** is based on 3 single insulated conductors that are rated up to 2000 volts and are directly buried. You will notice that only cables approved to be directly-buried are listed.

Table B.310.10 Ampacities of Three Single Insulated Conductors, Rated 0 Through 2000 Volts, Directly Buried in Earth Based on Ambient Earth Temperature of 20°C (68°F), Arrangement per Figure B.310.2, 100 Percent Load Factor, Thermal Resistance (Rho) of 90

Size (AWG or kcmil)	See Fig. B.310.2, Detail 9		See Fig. B.310.2, Detail 10		See Fig. B.310.2, Detail 9		See Fig. B.310.2, Detail 10		Size (AWG or kcmil)
	60°C (140°F)	75°C (167°F)	60°C (140°F)	75°C (167°F)	60°C (140°F)	75°C (167°F)	60°C (140°F)	75°C (167°F)	
	TYPES				TYPES				
	UF	USE	UF	USE	UF	USE	UF	USE	
	COPPER				ALUMINUM OR COPPER-CLAD ALUMINUM				
8	84	98	78	92	66	77	61	72	8
6	107	126	101	118	84	98	78	92	6
4	139	163	130	152	108	127	101	118	4
2	178	209	165	194	139	163	129	151	2
1	201	236	187	219	157	184	146	171	1
1/0	230	270	212	249	179	210	165	194	1/0
2/0	261	306	241	283	204	239	188	220	2/0
3/0	297	348	274	321	232	272	213	250	3/0
4/0	336	394	309	362	262	307	241	283	4/0
250	—	429	—	394	—	335	—	308	250
350	—	516	—	474	—	403	—	370	350
500	—	626	—	572	—	490	—	448	500
750	—	767	—	700	—	605	—	552	750
1000	—	887	—	808	—	706	—	642	1000
1250	—	979	—	891	—	787	—	716	1250
1500	—	1063	—	965	—	862	—	783	1500
1750	—	1133	—	1027	—	930	—	843	1750
2000	—	1195	—	1082	—	990	—	897	2000
Ambient Temp.(°C)	Correction Factors								Ambient Temp.(°F)
6–10	1.12	1.09	1.12	1.09	1.12	1.09	1.12	1.09	43–50
11–15	1.06	1.04	1.06	1.04	1.06	1.04	1.06	1.04	52–59
16–20	1.00	1.00	1.00	1.00	1.00	1.00	1.00	1.00	61–68
21–25	0.94	0.95	0.94	0.95	0.94	0.95	0.94	0.95	70–77
26–30	0.87	0.90	0.87	0.90	0.87	0.90	0.87	0.90	79–86

Reprinted with permission from NFPA 70, *National Electrical Code*®, Copyright ©2004, National Fire Protection Association, Quincy, MA 02169. This reprinted material is not the complete and official position of the National Fire Protection Association on the referenced subject which is represented only by the standard in its entirety.

Figures provided in this section, such as B.**[310.2]** and B.**[310.3]** below, illustrate cable installation requirements based on burial depths and quantity of conductors.

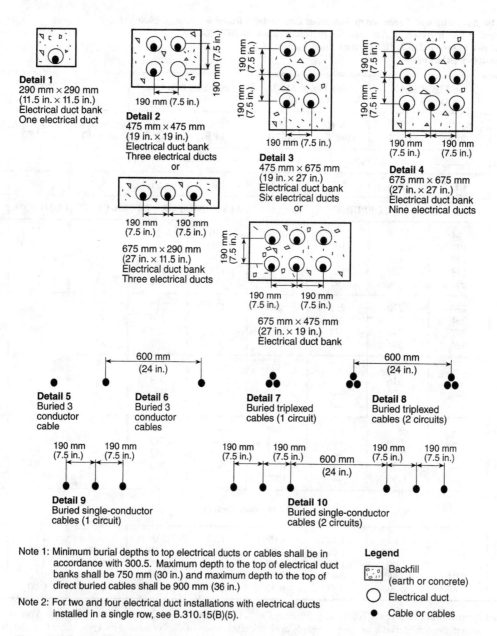

Figure B.310.2 Cable Installation Dimensions for Use with Table B.310.5 through Table B.310.10.

Reprinted with permission from NFPA 70, *National Electrical Code*®, Copyright ©2004, National Fire Protection Association, Quincy, MA 02169. This reprinted material is not the complete and official position of the National Fire Protection Association on the referenced subject which is represented only by the standard in its entirety.

Tables, Annexes, and Examples

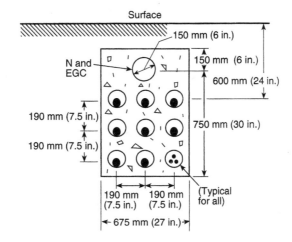

Design Criteria
Neutral and Equipment
 Grounding conductor (EGC)
 Duct = 150 mm (6 in.)
Phase Ducts = 75 to 125 mm (3 to 5 in.)
Conductor Material = Copper
Number of Cables per Duct = 3

Number of Cables per Phase = 9
Rho concrete = Rho Earth − 5

Rho PVC Duct = 650
Rho Cable Insulation = 500
Rho Cable Jacket = 650

Notes:
1. Neutral configuration per 300.5(I), Exception No. 2, for isolated phase installations in nonmagnetic ducts.
2. Phasing is A, B, C in rows or columns. Where magnetic electrical ducts are used, conductors are installed A, B, C per electrical duct with the neutral and all equipment grounding conductors in the same electrical duct. In this case, the 6-in. trade size neutral duct is eliminated.
3. Maximum harmonic loading on the neutral conductor cannot exceed 50 percent of the phase current for the ampacities shown in the table.
4. Metallic shields of Type MV-90 cable shall be grounded at one point only where using A, B, C phasing in rows or columns.

Size kcmil	TYPES RHW, THHW, THW, THWN, XHHW, USE, OR MV-90*			Size kcmil
	Total per Phase Ampere Rating			
	RHO EARTH 60 LF 50	RHO EARTH 90 LF 100	RHO EARTH 120 LF 100	
250	2340 (260A/Cable)	1530 (170A/Cable)	1395 (155A/Cable)	250
350	2790 (310A/Cable)	1800 (200A/Cable)	1665 (185A/Cable)	350
500	3375 (375A/Cable)	2160 (240A/Cable)	1980 (220A/Cable)	350

Ambient Temp. (°C)	For ambient temperatures other than 20°C (68°F), multiply the ampacities shown above by the appropriate factor shown below.					Ambient Temp. (°F)
6–10	1.09	1.09	1.09	1.09	1.09	43–50
11–15	1.04	1.04	1.04	1.04	1.04	52–59
16–20	1.00	1.00	1.00	1.00	1.00	61–68
21–25	0.95	0.95	0.95	0.95	0.95	70–77
26–30	0.90	0.90	0.90	0.90	0.90	79–86

*Limited to 75°C conductor temperature.

FPN Figure B.310.3 Ampacities of Single Insulated Conductors Rated 0 through 5000 Volts in Underground Electrical Ducts (Three Conductors per Electrical Duct), Nine Single-Conductor Cables per Phase Based on Ambient Earth Temperature of 20°C (68°F), Conductor Temperature 75°C (167°F).

Reprinted with permission from NFPA 70, *National Electrical Code*®, Copyright ©2004, National Fire Protection Association, Quincy, MA 02169. This reprinted material is not the complete and official position of the National Fire Protection Association on the referenced subject which is represented only by the standard in its entirety.

In Annex C you will find extensive conduit and tubing tables for conductors. For example, in Table C.1 you'll find that you are allowed to install three #8 AWG conductors in 1/2-inch EMT.

Table C.1 Maximum Number of Conductors or Fixture Wires in Electrical Metallic Tubing (EMT) (*Based on Table 1, Chapter 9*)

		CONDUCTORS								
	Conductor Size	Metric Designator (Trade Size)								
Type	(AWG kcmil)	16 (½)	21 (¾)	27 (1)	35 (1¼)	41 (1½)	53 (2)	63 (2½)	78 (3)	91 (3½)	103 (4)
RHH, RHW, RHW-2	14	4	7	11	20	27	46	80	120	157	201
	12	3	6	9	17	23	38	66	100	131	167
	10	2	5	8	13	18	30	53	81	105	135
	8	1	2	4	7	9	16	28	42	55	70
	6	1	1	3	5	8	13	22	34	44	56
	4	1	1	2	4	6	10	17	26	34	44
	3	1	1	1	4	5	9	15	23	30	38
	2	1	1	1	3	4	7	13	20	26	33
	1	0	1	1	1	3	5	9	13	17	22
	1/0	0	1	1	1	2	4	7	11	15	19
	2/0	0	1	1	1	2	4	6	10	13	17
	3/0	0	0	1	1	1	3	5	8	11	14
	4/0	0	0	1	1	1	3	5	7	9	12
	250	0	0	0	1	1	1	3	5	7	9
	300	0	0	0	1	1	1	3	5	6	8
	350	0	0	0	1	1	1	3	4	6	7
	400	0	0	0	1	1	1	2	4	5	7
	500	0	0	0	0	1	1	2	3	4	6
	600	0	0	0	0	1	1	1	3	4	5
	700	0	0	0	0	0	1	1	2	3	4
	750	0	0	0	0	0	1	1	2	3	4
	800	0	0	0	0	0	1	1	2	3	4
	900	0	0	0	0	0	1	1	1	3	3
	1000	0	0	0	0	0	1	1	1	2	3
	1250	0	0	0	0	0	0	1	1	1	2
	1500	0	0	0	0	0	0	1	1	1	1
	1750	0	0	0	0	0	0	1	1	1	1
	2000	0	0	0	0	0	0	1	1	1	1
TW	14	8	15	25	43	58	96	168	254	332	424
	12	6	11	19	33	45	74	129	195	255	326
	10	5	8	14	24	33	55	96	145	190	243
	8	2	5	8	13	18	30	53	81	105	135
RHH*, RHW*, RHW-2*, THHW, THW, THW-2	14	6	10	16	28	39	64	112	169	221	282
RHH*, RHW*, RHW-2*, THHW, THW	12	4	8	13	23	31	51	90	136	177	227
	10	3	6	10	18	24	40	70	106	138	177
RHH*, RHW*, RHW-2*, THHW, THW, THW-2	8	1	4	6	10	14	24	42	63	83	106

(*Continues*)

Tables, Annexes, and Examples

		CONDUCTORS									
	Conductor Size (AWG kcmil)	Metric Designator (Trade Size)									
Type		16 (½)	21 (¾)	27 (1)	35 (1¼)	41 (1½)	53 (2)	63 (2½)	78 (3)	91 (3½)	103 (4)
RHH*, RHW*, RHW-2*, TW, THW, THHW, THW-2	6	1	3	4	8	11	18	32	48	63	81
	4	1	1	3	6	8	13	24	36	47	60
	3	1	1	3	5	7	12	20	31	40	52
	2	1	1	2	4	6	10	17	26	34	44
	1	1	1	1	3	4	7	12	18	24	31
	1/0	0	1	1	2	3	6	10	16	20	26
	2/0	0	1	1	1	3	5	9	13	17	22
	3/0	0	1	1	1	2	4	7	11	15	19
	4/0	0	0	1	1	1	3	6	9	12	16
	250	0	0	1	1	1	3	5	7	10	13
	300	0	0	1	1	1	2	4	6	8	11
	350	0	0	0	1	1	1	4	6	7	10
	400	0	0	0	1	1	1	3	5	7	9
	500	0	0	0	1	1	1	3	4	6	7
	600	0	0	0	1	1	1	2	3	4	6
	700	0	0	0	0	1	1	1	3	4	5
	750	0	0	0	0	1	1	1	3	4	5
	800	0	0	0	0	1	1	1	3	3	5
	900	0	0	0	0	0	1	1	2	3	4
	1000	0	0	0	0	0	1	1	2	3	4
	1250	0	0	0	0	0	1	1	1	2	3
	1500	0	0	0	0	0	1	1	1	1	2
	1750	0	0	0	0	0	0	1	1	1	2
	2000	0	0	0	0	0	0	1	1	1	1
THHN, THWN, THWN-2	14	12	22	35	61	84	138	241	364	476	608
	12	9	16	26	45	61	101	176	266	347	443
	10	5	10	16	28	38	63	111	167	219	279
	8	3	6	9	16	22	36	64	96	126	161
	6	2	4	7	12	16	26	46	69	91	116
	4	1	2	4	7	10	16	28	43	56	71
	3	1	1	3	6	8	13	24	36	47	60
	2	1	1	3	5	7	11	20	30	40	51
	1	1	1	1	4	5	8	15	22	29	37
	1/0	1	1	1	3	4	7	12	19	25	32
	2/0	0	1	1	2	3	6	10	16	20	26
	3/0	0	1	1	1	3	5	8	13	17	22
	4/0	0	1	1	1	2	4	7	11	14	18
	250	0	0	1	1	1	3	6	9	11	15
	300	0	0	1	1	1	3	5	7	10	13
	350	0	0	1	1	1	2	4	6	9	11
	400	0	0	0	1	1	1	4	6	8	10
	500	0	0	0	1	1	1	3	5	6	8
	600	0	0	0	1	1	1	2	4	5	7
	700	0	0	0	1	1	1	2	3	4	6
	750	0	0	0	0	1	1	1	3	4	5
	800	0	0	0	0	1	1	1	3	4	5
	900	0	0	0	0	1	1	1	3	3	4
	1000	0	0	0	0	1	1	1	2	3	4
FEP, FEPB, PFA, PFAH, TFE	14	12	21	34	60	81	134	234	354	462	590
	12	9	15	25	43	59	98	171	258	337	430
	10	6	11	18	31	42	70	122	185	241	309
	8	3	6	10	18	24	40	70	106	138	177
	6	2	4	7	12	17	28	50	75	98	126
	4	1	3	5	9	12	20	35	53	69	88
	3	1	2	4	7	10	16	29	44	57	73
	2	1	1	3	6	8	13	24	36	47	60

(Continues)

Table C.1 *Continued*

		CONDUCTORS								
	Conductor Size (AWG kcmil)	Metric Designator (Trade Size)								
Type		16 (½)	21 (¾)	27 (1)	35 (1¼)	41 (1½)	53 (2)	63 (2½)	78 (3)	91 (3½)	103 (4)
PFA, PFAH, TFE	1	1	1	2	4	6	9	16	25	33	42
PFAH, TFE PFA, PFAH, TFE, Z	1/0	1	1	1	3	5	8	14	21	27	35
	2/0	0	1	1	3	4	6	11	17	22	29
	3/0	0	1	1	2	3	5	9	14	18	24
	4/0	0	1	1	1	2	4	8	11	15	19
Z	14	14	25	41	72	98	161	282	426	556	711
	12	10	18	29	51	69	114	200	302	394	504
	10	6	11	18	31	42	70	122	185	241	309
	8	4	7	11	20	27	44	77	117	153	195
	6	3	5	8	14	19	31	54	82	107	137
	4	1	3	5	9	13	21	37	56	74	94
	3	1	2	4	7	9	15	27	41	54	69
	2	1	1	3	6	8	13	22	34	45	57
	1	1	1	2	4	6	10	18	28	36	46
XHH, XHHW, XHHW-2, ZW	14	8	15	25	43	58	96	168	254	332	424
	12	6	11	19	33	45	74	129	195	255	326
	10	5	8	14	24	33	55	96	145	190	243
	8	2	5	8	13	18	30	53	81	105	135
	6	1	3	6	10	14	22	39	60	78	100
	4	1	2	4	7	10	16	28	43	56	72
	3	1	1	3	6	8	14	24	36	48	61
	2	1	1	3	5	7	11	20	31	40	51
XHH, XHHW, XHHW-2	1	1	1	1	4	5	8	15	23	30	38
	1/0	1	1	1	3	4	7	13	19	25	32
	2/0	0	1	1	2	3	6	10	16	21	27
	3/0	0	1	1	1	3	5	9	13	17	22
	4/0	0	1	1	1	2	4	7	11	14	18
	250	0	0	1	1	1	3	6	9	12	15
	300	0	0	1	1	1	3	5	8	10	13
	350	0	0	1	1	1	2	4	7	9	11
	400	0	0	0	1	1	1	4	6	8	10
	500	0	0	0	1	1	1	3	5	6	8
	600	0	0	0	1	1	1	2	4	5	6
	700	0	0	0	0	1	1	2	3	4	6
	750	0	0	0	0	1	1	1	3	4	5
	800	0	0	0	0	1	1	1	3	4	5
	900	0	0	0	0	1	1	1	3	3	4
	1000	0	0	0	0	0	1	1	2	3	4
	1250	0	0	0	0	0	1	1	1	2	3
	1500	0	0	0	0	0	1	1	1	1	3
	1750	0	0	0	0	0	0	1	1	1	2
	2000	0	0	0	0	0	0	1	1	1	1

(Continues)

Table C.1 *Continued*

		FIXTURE WIRES					
	Conductor Size (AWG/ kcmil)	Metric Designator (Trade Size)					
Type		16 (½)	21 (¾)	27 (1)	35 (1¼)	41 (1½)	53 (2)
FFH-2, RFH-2, RFHH-3	18	8	14	24	41	56	92
	16	7	12	20	34	47	78
SF-2, SFF-2	18	10	18	30	52	71	116
	16	8	15	25	43	58	96
	14	7	12	20	34	47	78
SF-1, SFF-1	18	18	33	53	92	125	206
RFH-1, RFHH-2, TF, TFF, XF, XFF	18	14	24	39	68	92	152
RFHH-2, TF, TFF, XF, XFF	16	11	19	31	55	74	123
XF, XFF	14	8	15	25	43	58	96
TFN, TFFN	18	22	38	63	108	148	244
	16	17	29	48	83	113	186
PF, PFF, PGF, PGFF, PAF, PTF, PTFF, PAFF	18	21	36	59	103	140	231
	16	16	28	46	79	108	179
	14	12	21	34	60	81	134
ZF, ZFF, ZHF, HF, HFF	18	27	47	77	133	181	298
	16	20	35	56	98	133	220
	14	14	25	41	72	98	161
KF-2, KFF-2	18	39	69	111	193	262	433
	16	27	48	78	136	185	305
	14	19	33	54	93	127	209
	12	13	23	37	64	87	144
	10	8	15	25	43	58	96
KF-1, KFF-1	18	46	82	133	230	313	516
	16	33	57	93	161	220	362
	14	22	38	63	108	148	244
	12	14	25	41	72	98	161
	10	9	16	27	47	64	105
XF, XFF	12	4	8	13	23	31	51
	10	3	6	10	18	24	40

Notes:

1. This table is for concentric stranded conductors only. For compact stranded conductors, Table C.1(A) should be used.

2. Two-hour fire-rated RHH cable has ceramifiable insulation which has much larger diameters than other RHH wires. Consult manufacturer's conduit fill tables.

*Types RHH, RHW, and RHW-2 without outer covering.

Reprinted with permission from NFPA 70, *National Electrical Code*®, Copyright ©2004, National Fire Protection Association, Quincy, MA 02169. This reprinted material is not the complete and official position of the National Fire Protection Association on the referenced subject which is represented only by the standard in its entirety.

Table C.**[1(A)]** provides similar information for compact conductors.

Table C.1(A) Maximum Number of Compact Conductors in Electrical Metallic Tubing (EMT)
(Based on Table 1, Chapter 9)

		COMPACT CONDUCTORS								
	Conductor Size (AWG/ kcmil)	Metric Designator (Trade Size)								
Type		16 (½)	21 (¾)	27 (1)	35 (1¼)	41 (1½)	53 (2)	63 (2½)	78 (3)	91 (3½)	103 (4)
THW, THW-2, THHW	8	2	4	6	11	16	26	46	69	90	115
	6	1	3	5	9	12	20	35	53	70	89
	4	1	2	4	6	9	15	26	40	52	67
	2	1	1	3	5	7	11	19	29	38	49
	1	1	1	1	3	4	8	13	21	27	34
	1/0	1	1	1	3	4	7	12	18	23	30
	2/0	0	1	1	2	3	5	10	15	20	25
	3/0	0	1	1	1	3	5	8	13	17	21
	4/0	0	1	1	1	2	4	7	11	14	18
	250	0	0	1	1	1	3	5	8	11	14
	300	0	0	1	1	1	3	5	7	9	12
	350	0	0	1	1	1	2	4	6	8	11
	400	0	0	0	1	1	1	4	6	8	10
	500	0	0	0	1	1	1	3	5	6	8
	600	0	0	0	1	1	1	2	4	5	7
	700	0	0	0	1	1	1	2	3	4	6
	750	0	0	0	0	1	1	1	3	4	5
	900	0	0	0	0	1	1	2	3	4	5
	1000	0	0	0	0	1	1	1	2	3	4
THHN, THWN, THWN-2	8	—	—	—	—	—	—	—	—	—	—
	6	2	4	7	13	18	29	52	78	102	130
	4	1	3	4	8	11	18	32	48	63	81
	2	1	1	3	6	8	13	23	34	45	58
	1	1	1	2	4	6	10	17	26	34	43
	1/0	1	1	1	3	5	8	14	22	29	37
	2/0	1	1	1	3	4	7	12	18	24	30
	3/0	0	1	1	2	3	6	10	15	20	25
	4/0	0	1	1	1	3	5	8	12	16	21
	250	0	1	1	1	1	4	6	10	13	16
	300	0	0	1	1	1	3	5	8	11	14
	350	0	0	1	1	1	3	5	7	10	12
	400	0	0	1	1	1	2	4	6	9	11
	500	0	0	0	1	1	1	4	5	7	9
	600	0	0	0	1	1	1	3	4	6	7
	700	0	0	0	1	1	1	2	4	5	7
	750	0	0	0	1	1	1	2	4	5	6
	900	0	0	0	0	1	1	2	3	4	5
	1000	0	0	0	0	1	1	1	3	3	4
XHHW, XHHW-2	8	3	5	8	15	20	34	59	90	117	149
	6	1	4	6	11	15	25	44	66	87	111
	4	1	3	4	8	11	18	32	48	63	81
	2	1	1	3	6	8	13	23	34	45	58
	1	1	1	2	4	6	10	17	26	34	43
	1/0	1	1	1	3	5	8	14	22	29	37
	2/0	1	1	1	3	4	7	12	18	24	31
	3/0	0	1	1	2	3	6	10	15	20	25
	4/0	0	1	1	1	3	5	8	13	17	21
	250	0	1	1	1	2	4	7	10	13	17
	300	0	0	1	1	1	3	6	9	11	14
	350	0	0	1	1	1	3	5	8	10	13
	400	0	0	1	1	1	2	4	7	9	11
	500	0	0	0	1	1	1	4	6	7	9
	600	0	0	0	1	1	1	3	4	6	8
	700	0	0	0	1	1	1	2	4	5	7
	750	0	0	0	1	1	1	2	3	5	6
	900	0	0	0	0	1	1	2	3	4	5
	1000	0	0	0	0	1	1	1	3	4	5

Definition: *Compact stranding* is the result of a manufacturing process where the standard conductor is compressed to the extent that the interstices (voids between strand wires) are virtually eliminated.

Reprinted with permission from NFPA 70, *National Electrical Code*®, Copyright ©2004, National Fire Protection Association, Quincy, MA 02169. This reprinted material is not the complete and official position of the National Fire Protection Association on the referenced subject which is represented only by the standard in its entirety.

If you run the EMT in a straight line, the interior diameter is 0.622 inches, but as soon as you include a bend in the conduit you create a situation where the major internal diameter to the EMT can actually become slightly larger, and therefore one of the conductors could slip between the other two. This creates the potential for a jam where the conductors exit the bend. Those of you sitting for the Master Electrician exam should note that the formula to calculate jam ratio is:

$$\text{Jam Ratio} = \frac{\text{ID of the raceway}}{\text{OD of the conductor}}$$

To avoid a possible jam within the tubing, use the recommended jam ratio of 2.8 and 3.2. For various combinations of conductors of different sizes, you need to look at Tables **[5]** and **[5.A]**, which provide the dimensions of conductors, and then Table **[4]** for the applicable tubing or conduit dimensions.

Annex D is a section of detailed examples using math formulas for such applications as general lighting loads, calculating the size for neutrals for feeders, demand factors, calculated loads for service conductors, and more. The examples vary from single family units to multi-family dwellings, as well as electric range loads and motor circuit calculations. Figure **[D9]** below illustrates an adjustable speed drive control and how to determine the feeder ampacity. Figure **[D10]** illustrates an adjustable speed drive control from the example provided in **[D10]**.

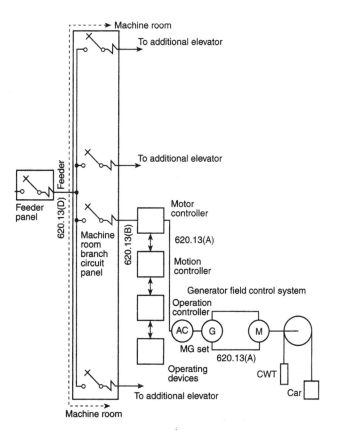

Figure D9 Generator Field Control.

Reprinted with permission from NFPA 70, *National Electrical Code*®, Copyright ©2004, National Fire Protection Association, Quincy, MA 02169. This reprinted material is not the complete and official position of the National Fire Protection Association on the referenced subject which is represented only by the standard in its entirety.

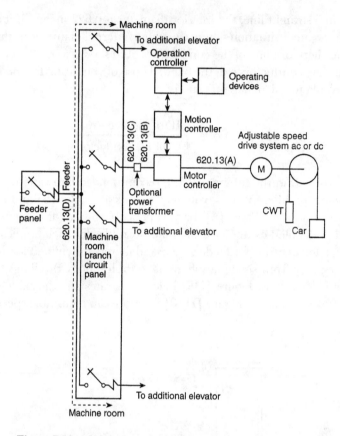

Figure D10 Adjustable Speed Drive Control.

Reprinted with permission from NFPA 70, *National Electrical Code®*, Copyright ©2004, National Fire Protection Association, Quincy, MA 02169. This reprinted material is not the complete and official position of the National Fire Protection Association on the referenced subject which is represented only by the standard in its entirety.

Annex E describes five different types of construction, including fire-resistant and non-rated materials, and provides two tables. Annex F is particularly handy, not only because is compares the location of code articles over the last three publications, but because it briefly describes the articles so that you can jump to that section of the book if you are struggling to find an answer. Finally, Annex G seeks to explain how the code is administered and by whom, as well as the issuing of permits, consequences for violations of the code, penalties, and the process for appeals.

We'll take a moment here to review some of the information in Chapter nine in the form of an informational quiz. Although this information is not required as part of the code, it will help you answer questions related to the subjects listed in the tables and annexes.

Table E.1 Fire Resistance Ratings (in hours) for Type I through Type V Construction

	Type I		Type II			Type III		Type IV	Type V	
	443	332	222	111	000	211	200	2HH	111	000
Exterior Bearing Walls –										
Supporting more than one floor, columns, or other bearing walls......	4	3	2	1	0[1]	2	2	2	1	0[1]
Supporting one floor only.............	4	3	2	1	0[1]	2	2	2	1	0[1]
Supporting a roof only.................	4	3	1	1	0[1]	2	2	2	1	0[1]
Interior Bearing Walls –										
Supporting more than one floor, columns, or other bearing walls......	4	3	2	1	0	1	0	2	1	0
Supporting one floor only.............	3	2	2	1	0	1	0	1	1	0
Supporting roofs only..................	3	2	1	1	0	1	0	1	1	0
Columns –										
Supporting more than one floor, columns, or other bearing walls......	4	3	2	1	0	1	0	H[2]	1	0
Supporting one floor only.............	3	2	2	1	0	1	0	H[2]	1	0
Supporting roofs only..................	3	2	1	1	0	1	0	H[2]	1	0
Beams, Girders, Trusses & Arches –										
Supporting more than one floor, columns, or other bearing walls......	4	3	2	1	0	1	0	H[2]	1	0
Supporting one floor only.............	3	2	2	1	0	1	0	H[2]	1	0
Supporting roofs only..................	3	2	1	1	0	1	0	H[2]	1	0
Floor Construction	3	2	2	1	0	1	0	H[2]	1	0
Roof Construction	2	1½	1	1	0	1	0	H[2]	1	0
Exterior Nonbearing Walls[3]	0[1]	0[1]	0[1]	0[1]	0[1]	0[1]	0[1]	0[1]	0[1]	0[1]

▓ Those members that shall be permitted to be of approved combustible material.

[1] See A-3-1 in NFPA 220.
[2] "H" indicates heavy timber members; see text for requirements.
[3] Exterior nonbearing walls meeting the conditions of acceptance of NFPA 285, *Standard Method of Test for the Evaluation of Flammability Characteristics of Exterior Non-Load-Bearing Wall Assemblies Containing Combustible Components Using the Intermediate-Scale, Multistory Test Apparatus*, shall be permitted to be used.

Source: Table 3.1 from NFPA 220, *Standard on Building Construction*, 1999.

Reprinted with permission from NFPA 70, *National Electrical Code*®, Copyright ©2004, National Fire Protection Association, Quincy, MA 02169. This reprinted material is not the complete and official position of the National Fire Protection Association on the referenced subject which is represented only by the standard in its entirety.

QUIZ 1—REVIEW

1. The percentage area for one wire running in 2-inch RNC schedule 40 conduit is which of the following:

   a. 1126 square inches
   b. 31 percent
   c. 3.291 inches
   d. 1.744 square inches

2. Three conductors with a total combined cross-sectional value of 0.0633 shall be run in which of the following sizes of flexible metal conduit:

   a. 2 inch
   b. 1 1/2 inch
   c. 1 1/4 inch
   d. 1 inch

3. The size of rigid metal conduit (RMC) required to enclose 18 #10 THWN copper conductors is which of the following:

   a. 1 1/2 inch
   b. 1 1/4 inch
   c. 1 inch
   d. 3/4 inch

4. In an installation containing three 350kcmil THWN copper conductors in 2 1/2-inch EMT, the conductors may become jammed.

   a. True
   b. False

5. The Code requires which of the following EMT nipple sizes to enclose 28 #12 THWN copper conductors installed between a panelboard and junction box:

   a. 1 1/2 inch
   b. 1 1/4 inch
   c. 1 inch
   d. 3/4 inch

6. **MASTER LEVEL QUESTION:** The size RMC required to enclose a 1 5/8-inch (1.625 diameter) multiconductor cable is which of the following:

   a. 1 1/2 inches
   b. 2 inches
   c. 2 1/2 inches
   d. 3 inches

7. To enclose two #6 THWN copper conductors, two #8 THWN copper conductors, and four #10 THWN copper conductors, which of the following sizes of electrical metallic tubing (EMT) is required:

   a. 1 inch
   b. 1 1/4 inches
   c. 1 1/2 inches
   d. None of the above

Tables, Annexes, and Examples **237**

8. Rho stands for which of the following:

    a. Resistive Hard Overcurrent   **b.** Reactive Current Overall

    c. Temperature Rise Factor   **d.** Soil Thermal Resistivity

9. The maximum number of #8 AWG THHN conductors permitted in 1-inch EMT conduit is which of the following:

    a. 1   **b.** 2

    c. 3   **d.** None of the above

10. Imagine a single-family dwelling with the following characteristics: 1500 square feet, a 12 kW electric range, a 5.5 kW, 240 volt electric dryer, one 6-amp, 240 volt electric room air conditioner and one 12-amp 115 volt room air conditioner, an 8-amp 115 volt garbage disposal, and a 10-amp 120 volt dishwasher. Assume that the nameplates list 115 volts and 230 volts for use with 120 volt and 240 volt nominal systems. The service size for this dwelling unit would be which of the following:

    a. 100 amp   **b.** 110 amp

    c. 200 amp   **d.** 240 amp

TESTING

Now that you have had the opportunity to study and review all nine chapters of the NEC® it is time for you to apply your knowledge and testing abilities to some comprehensive sample tests. These are not timed tests and you are allowed to use your NEC® book to look up or confirm your answers. You must get 35 questions correct in order to score a 70 percent, which is the typical requirement to pass your licensing exam.

QUIZ 2

1. If an outlet is in an underfloor raceway and is disconnected, then which of the following actions must be taken on the circuit conductors supplying the outlet:

    a. A blank cover plate installed on the outlet box

    b. Be removed from the raceway

    c. Be spliced and reconnected

    d. None of the above

2. Given a single-family dwelling with a 3 feet wide front door, an attached garage on the south side of the dwelling with an 8 feet wide door, and a back entrance door which is 2 feet 6 inches wide, the total number of lighting outlets required for the dwelling is which of the following:

    a. 1
    b. 2
    c. 3
    d. 4

3. Other than service conductors, conductors shall not be installed in the same service raceway unless they are which of the following:

    a. Grounding conductors

    b. Shielded conductors

    c. Both of the above

    d. Either of the above

4. A space is considered a bathroom if it contains a toilet and a tub or shower.

    a. True
    b. False

5. Conductor sizes are listed in AWG and by which of the following standard types:

    a. Insulation value
    b. CM
    c. DC
    d. All of the above

6. A dwelling with a three-wire, 240/120 volt single-phase 200 amp service requires which of the following conductor sizes:

    a. 18 AWG
    b. #1/0 THW
    c. #2/0 THW
    d. All of the above

7. Overcurrent devices in dwelling units and hotel guest rooms shall not be located in which of the following areas:
   a. Bedrooms
   b. Bathrooms
   c. Kitchens
   d. All of the above

8. Disregarding the outer covering, the area in square inches for a 10 feet section of #14 RHW is which of the following:
   a. 20.9
   b. 2.090
   c. .0209
   d. 1.40

9. All conductors in a multiwire branch circuit shall originate from the same feeder.
   a. True
   b. False

10. Nails used to mount knobs shall not be smaller than 16 penny.
    a. True
    b. False

11. In order to balance a 3-wire, single-phase 230/115 volt circuit, which of the following must be done:
    a. The neutral conductor must carry the unbalanced current.
    b. The hot conductor must carry the total current.
    c. The neutral conductor should be used for grounding only.
    d. None of the above

12. In fixed electric space heating equipment, a disconnection means shall be provided from all ungrounded conductors to which of the following equipment components:
    a. The heater
    b. Motor controllers
    c. Supplementary overcurrent protection devices
    d. All of the above

13. If an AC system operates at less than 1000 volts and is grounded at any point then the grounded conductor shall be run to each service.
    a. True
    b. False

14. Computation of fluorescent fixture loads shall be based on which of the following:
    a. The total ampere rating of the fixture
    b. The total wattage of the lamps
    c. The total volts of the wiring system
    d. None of the above

15. A multiwire branch circuit may be used to supply which of the following:
    a. A 120/240 volt system to one piece of utilization equipment
    b. A 120/240 volt system where all conductors are opened simultaneously
    c. Both of the above
    d. None of the above

16. If nonmetallic sheathed cable is installed in an attic that is not accessible by stairs or a permanent ladder, then the cable must be protected within which of the following:
    a. 2 feet of any support
    b. 6 feet of a scuttle hole
    c. 10 feet of any connection
    d. All of the above

17. Flexible cords must be secured underneath showcases so that the wiring will not be exposed to mechanical damage.
    a. True
    b. False

18. Parallel conductors in each phase or neutral shall be which of the following:
    a. The same length and terminated in the same manner
    b. Any size trade wire
    c. Either A or B
    d. None of the above

19. If single conductors are not installed in a raceway and stacked longer than 2 feet without maintaining spacing, then which of the following is true:
    a. The ampacity of each conductor shall be increased
    b. The ampacity of each conductor shall be reduced
    c. The conductor must increase to the next size up
    d. None of the above

20. Temporary power installations for a 120/240 volt Christmas decoration display shall not remain in place longer than 60 days.
    a. True
    b. False

Tables, Annexes, and Examples

21. If one end of a #6 copper conductor is bonded to the service raceway or equipment with 6 inches or more of the other end accessible on the outside wall of a dwelling unit, then this would be which of the following:

   a. An illegal installation

   b. An example of an approved means for external connection of a bonding conductor to the service raceway

   c. An example of a permitted service entrance cable installation

   d. None of the above

22. The maximum number of overcurrent devices permitted in a lighting panel is which of the following:

   a. 10
   b. 24
   c. 42
   d. None of the above

23. Which of the following is true of open conductors on insulators:

   a. They must be supported every 2 feet.

   b. They must be waterproof.

   c. They must be size #16 AWG or larger.

   d. They must be covered if they are within 10 feet of a building.

24. For general applications, a nominal voltage of 120/240 is used to compute the ampere load of a conductor.

   a. True
   b. False

25. Certain conductors are permitted to be installed where the operating temperature in the installation location exceeds the rating of the conductor insulation.

   a. True
   b. False

26. Open conductors shall be supported on which of the following:

   I. Glass or porcelain knobs

   II. Racks or brackets

   III. Strain insulators

   a. I or III
   b. I or II
   c. II or III
   d. Any of the above

27. NM cable installations shall conform to which of the following:
  a. It may be covered with plaster
  b. It may be fished through air voids in masonry block or tile walls
  c. Both of the above
  d. None of the above

28. The ampacity of multiconductor cables installed in a cable tray shall be derated based on which of the following:
  a. The total number of current carrying conductors in the cable
  b. The total number of current carrying conductors in the cable tray
  c. Both of the above
  d. None of the above

29. Switching should be done in which of the following:
  a. The wall
  b. The ungrounded conductor
  c. The grounded conductor
  d. A termination box

30. Resistors and reactors over 600 volts shall not be installed within close proximity of a combustible material that could constitute a fire hazard.
  a. True
  b. False

31. Resistors and reactors over 600 volts shall have a minimum clearance from combustible materials of which of the following:
  a. 3 inches
  b. 6 inches
  c. 12 inches
  d. 18 inches

32. Turning vanes, pressure plates or other devices may need to be installed on the inlet side of a duct heater if the heater installation is within ____ of an outlet for a heat pump, air conditioning elbows, baffle plates or other obstructions in duct work:
  a. 1 foot
  b. 2 feet
  c. 3 feet
  d. 4 feet

33. If an indoor transformer is insulated with dielectric fluid, it shall be installed in a vault if the transformer is rated over which of the following:
  a. 35 kVA
  b. 35 kv
  c. 3500 VA
  d. None of the above

Tables, Annexes, and Examples

34. Fuses must be clearly marked with which of the following:

    a. Voltage rating

    b. Ampere rating

    c. Interruption rating, if other than 10,000 amperes

    d. All of the above

35. If an installation requires a 10.25 cubic inch device box, then the minimum size of the box would be which of the following:

    a. 3 inches × 2inches × 2 1/4inches

    b. 3inches × 2inches × 2inches

    c. 2 1/4inches × 2 1/4inches × 2inches

    d. 2inches × 2inches × 3inches

36. If a new office building has two service heads and will have only one service drop, then the maximum distance permitted between the service heads shall be which of the following:

    a. 6 feet   **b.** 5 feet

    c. 4 feet   **d.** 3 feet

37. From a service lateral to a branch circuit limited load, the minimum size conductor permitted is which of the following:

    a. #6 copper   **b.** #8 copper

    c. #10 copper   **d.** None of the above

38. Type FCC cables shall be clearly and permanently marked with which of the following:

    a. The conductor material

    b. The ampacity

    c. The maximum temperature rating

    d. All of the above

39. The optional method of calculation shall be permitted for a multifamily dwelling unit if none of the units is supplied by more than one feeder.

    a. True   **b.** False

40. Control and operating circuits for X-ray and auxiliary equipment protected by not larger than a 20 ampere overcurrent device shall be permitted to have which of the following:

    a. Size #18 fixture wires   **b.** Size #16 fixture wires

    c. Both of the above   **d.** None of the above

41. Under no condition shall the distance between nonmetallic wireway supports exceed which of the following:

- **a.** 10 feet
- **b.** 8 feet
- **c.** 6 feet
- **d.** 36 inches

42. Grounded interior wiring shall not be permitted to be connected to a power supply unless the supply system contains a corresponding conductor that is which of the following:

- **a.** Ungrounded
- **b.** Grounded
- **c.** Bonded
- **d.** None of the above

43. Flexible cords shall be secured to the underside of a group of showcases so that the free lead at the end of the showcases has a female fitting that extends no less than 3 inches beyond the case.

- **a.** True
- **b.** False

44. A grounded conductor that is #1100 kcmil or less brought to the service shall comply with which of the following:

- **a.** Shall not be larger than the minimum size of the grounding electrode
- **b.** Shall not be smaller than the minimum size of the grounding electrode
- **c.** Shall be 2 times the minimum size of the grounding electrode
- **d.** None of the above

45. Even if an outlet box is listed and marked by the manufacturer as suitable to be used as the sole support for a fixture, the fixture shall not weigh more than which of the following:

- **a.** 100 pounds
- **b.** 70 pounds
- **c.** 50 pounds
- **d.** 35 pounds

46. The ampacity of IGS cable #250 kcmil is which of the following:

- **a.** 119 amperes
- **b.** 114 amperes
- **c.** 125 amperes
- **d.** 500 amperes

47. If the input current to an autotransformer is less than 9 amps, then the maximum overcurrent protection shall be which of the following:

- **a.** 300 percent of the input current
- **b.** 187 percent of the input current
- **c.** 167 percent of the input current
- **d.** 110 percent of the input current

48. Strut-type channel raceway shall be secured at which of the following intervals:

   I. No farther than 10 feet from every outlet box

   II. No farther than 3 feet from every outlet box

   III. Every 10 feet or less

   IV. Every 6 feet or less

   a. I & II
   b. I & III
   c. II & III
   d. II & IV

49. A moisture seal is required at all points of termination for mineral-insulated cable.

   a. True
   b. False

50. A 20 ampere rated branch circuit in the living room of a dwelling unit shall be permitted to carry a maximum load of which of the following:

   a. 10 amps
   b. 15 amps
   c. 20 amps
   d. 40 amps

QUIZ 3

1. If an electrician must stand on a 6 feet extension ladder in order to reach a lighting fixture, then the junction box for the fixtures is considered to be which of the following:

   a. Readily accessible
   b. Easily accessible
   c. Accessible
   d. Concealed

2. All branch circuits that supply the 125 volt 15 amp and 20 amp outlets of a dwelling unit bedroom shall be protected by an arc-flash device rated for the branch circuit.

   a. True
   b. False

3. An automatic overcurrent device that protects service conductors supplying a specific load, such as an electric water heater, shall be permitted to be locked or sealed in order to prevent which of the following:

   a. Corrosion
   b. Tampering
   c. Derating of the conductors
   d. Tripping

4. A service mast used to support service drop conductors shall be supported by which of the following:
   a. Glass or porcelain
   b. Braces or guys
   c. Studs or hangers
   d. None of the above

5. A transformer enclosure that extends directly to an underwater pool light and forms a shell must have which of the following:
   a. One grounding terminal
   b. Be of a nonmetallic material
   c. The number of conduit entries plus one
   d. None of the above

6. Conductors for festoon lighting shall be supported by messenger wire if the lighting run exceeds which of the following:
   a. 70 pounds
   b. 40 feet
   c. 100 volts
   d. None of the above

7. The area in square inches of a #8 bare conductor in a raceway is equal to which of the following:
   a. 0.778 square inches
   b. 0.012 square inches
   c. 0.138 square inches
   d. None of the above

8. A household appliance with surface heating elements that have a maximum demand of more than 60 amperes shall have which of the following:
   a. Power supply that is subdivided into two or more circuits
   b. Overcurrent protection rated at not over 50 amperes
   c. Both of the above
   d. None of the above

9. If three 1/2 horsepower motors are located on a nominal 120 volt branch circuit, then the branch circuit shall be protected at which of the following levels:
   a. Not over 15 amps
   b. Not less than 15 amps
   c. Not over 20 amps
   d. Not over 100 volts

10. Swimming pool light fixtures shall not be installed on supply circuits that are over which of the following:
    a. 150 volts between conductors
    b. 120 volts
    c. 40 feet in length
    d. None of the above

Tables, Annexes, and Examples

11. Messenger supported wiring shall not be used for which of the following:

    a. In hoistways

    b. Where subjected to severe physical damage

    c. Both of the above

    d. None of the above

12. Which of the following entrances does not require a switched outlet:

    a. An attic

    b. A drive through garage door

    c. A walk through porch door

    d. All of these require a switched outlet

13. An insulated conductor installed as a grounded conductor of circuits shall have an outer identification of white or gray and shall not be smaller than which of the following:

    a. #2 AWG   **b.** # 4 AWG

    c. #6 AWG   **d.** #1/0 AWG

14. An indoor dry-type autotransformer rated at 112 kVA shall have a separation of at least 18 inches from any combustible material.

    a. True   **b.** False

15. All joints, splices and free ends of conductors shall have which of the following:

    a. Insulation equivalent to the conductor

    b. Insulation as thick as the conductor

    c. Rubber insulation

    d. All of the above

16. All heating elements that are part of an appliance shall be marked with the voltage and amperes rating, or the voltage and wattage rating, or with the manufacturer's part number if the elements are which of the following:

    a. Rated over 1 ampere   **b.** Replaceable in the field

    c. Both of the above   **d.** None of the above

17. Open motors with communicators shall be located so that any sparks cannot reach any combustible materials; however which of the following is also true:

   a. This applies only to motors over 600 volts

   b. This requirement shall not prohibit the motors from being on wooden floors

   c. This does not prohibit these motors from being located in a Class 1 area

   d. All of the above

18. Cord-and-plug connections are permitted for a range hoods if they are connected with a flexible cord identified as suitable for this use as long as which of the following is true:

   a. The cord is not less than 18inches long

   b. The cord is not more than 36inches long

   c. Both of the above

   d. None of the above

19. Which of the following receptacles do not need to be grounded:

   a. Receptacles located on an interior garage wall

   b. Receptacles for electric ranges

   c. Outdoor receptacles

   d. All of the above

20. Circuits for a show window are calculated at which of the following:

   a. 120 volts per square foot

   b. 200 va per linear foot

   c. 1500 watts per foot

   d. None of the above

21. A bare conductor feeder requires a moisture seal at all points of termination.

   a. True   **b.** False

22. A feeder supplies a residential electric range and electric clothes dryer, and the maximum unbalanced load on the neutral conductor is considered to be which of the following percentages of the load on the ungrounded conductors:

   a. 20 percent   **b.** 50 percent

   c. 70 percent   **d.** 85 percent

23. The AWG number of a conductor varies from the ampacity in which of the following manners:

    a. Directly   **b.** Proportionately

    c. Inversely   **d.** In no way

24. Disagreements between an electrical Inspector and a Master electrician shall be settled by which of the following:

    a. The local authority having jurisdiction

    b. The state electrical board

    c. Occupational Safety and Health Administration

    d. Civil court at the local level

25. On an electric motor, "Z.P." is the abbreviated marking for which of the following:

    a. Impedance Protected   **b.** Zero Polarity

    c. Zero Phase   **d.** None of the above

26. The equipment grounding conductors of a derived system shall be connected to a bonding jumper at any point on the separately derived system from the source to the first system disconnect.

    a. True   **b.** False

27. In a lighting panel, the maximum number of overcurrent devices permitted is which of the following:

    a. 52   **b.** 48

    c. 42   **d.** 24

28. Type FCC cable wiring systems shall be installed in which of the following:

    a. Under carpets   **b.** In vertical riser

    c. Along tile   **d.** None of the above

29. All of the conductors in a multiwire branch circuit must meet which of the following requirements:

    a. Originate from the same panelboard

    b. Terminate at the same location

    c. Originate from the same feeder

    d. None of the above

30. A receptacle mounted on a portable generator does not have to be grounded.

    a. True   **b.** False

31. Which of the following identifying items of information is not required on a motor nameplate:
    a. Voltage
    b. Horsepower
    c. Watts
    d. Manufacturer's identification

32. A single receptacle shall have a rating of which of the following:
    a. 50 percent of the branch circuit rating
    b. 100 percent of the branch circuit rating
    c. 15 amps
    d. None of the above

33. Safe access in the form of a permanent ladder shall be provided to the working spaces around electric equipment over 600 volts that is installed on or in which of the following:
    a. Mezzanine floor
    b. Attic
    c. Balconies
    d. All of the above

34. A multi-outlet assembly shall be permitted in a hoistway.
    a. True
    b. False

35. Parallel conductors in each phase or neutral shall meet which of the following requirements:
    a. Be the same length and be terminated in the same manner
    b. Be the same AWG size
    c. Be the same amperage
    d. All of the above

36. Surge arresters shall meet which of the following requirements:
    a. May be located outdoors
    b. Shall be made inaccessible to unqualified persons
    c. May be located indoors
    d. All of the above

37. A general-use snap switch shall only be used to control which of the following on alternating current circuits:

   a. Resistive loads that do not exceed the rating of the switch

   b. Motor loads not in excess of 80% of the switch ampere rating

   c. Inductive loads not exceeding the ampere rating of the switch

   d. All of the above

38. Open land subject to cultivation has a clearance requirement for 6900 volt to ground conductors of which of the following:

   a. 10 feet
   b. 18 1/2 feet
   c. 20 feet
   d. 32 feet

39. A device intended to interrupt under standard use conditions has a maximum current-at-rated voltage that is defined as which of the following:

   a. The overload current

   b. The interrupting rating

   c. The ground fault interrupter load

   d. None of the above

40. Receptacles shall be of the grounding type if they are installed on which of the following:

   a. 30 amp circuits

   b. 15 or 20 amp branch circuits

   c. 20 amp circuits

   d. None of the above

41. A controller that is not in sight of a motor location shall meet which of the following requirements:

   a. Be capable of being shut off

   b. Be connected to a bonding jumper

   c. Be capable of being locked in the open position

   d. Be capable of being set down in the off position

42. The service conductors that run between the street main and the first point of connection to an underground service entrance are which of the following:

   a. The service lateral

   b. A grounded loop system

   c. A service drop

   d. None of the above

43. Excluding any exceptions, a single family dwelling unit shall have which of the following sized branch circuit ratings for a 10kW electric range:

- **a.** At least 30 amps
- **b.** At least 40 amps
- **c.** 20 amps
- **d.** 50 amps

44. Fluorescent light fixtures supported independently of an outlet box shall meet which of the following requirements:

- **a.** May be connected by a metal raceway
- **b.** May use nonmetallic sheathed romex cable
- **c.** May be connected by a nonmetallic raceway
- **d.** All of the above

45. Service entrance equipment includes which of the following:
   I. Panelboards
   II. Meter enclosures
   III. Service disconnecting means

- **a.** I only
- **b.** II & III only
- **c.** I & III only
- **d.** II only

46. An autotransformer that is 600 volts or less shall be protected by an individual overcurrent protection device, which is installed in series with each ungrounded conductor.

- **a.** True
- **b.** False

47. Medium voltage cable insulation is rated for which of the following:

- **a.** 600 volts and higher
- **b.** 600 volts and lower
- **c.** 2001 volts and higher
- **d.** None of the above

48. The minimum size wire for a fixture rated at 7 amps is which of the following:

- **a.** #12 AWG
- **b.** #14 AWG
- **c.** #18 AWG
- **d.** None of the above

49. A 2400 volt lead cable shall be permitted to be bent in accordance with which of the following:

- **a.** A total of 4 times
- **b.** Up to 12 times its diameter
- **c.** Using up to three 90 degree angles
- **d.** None of the above

Tables, Annexes, and Examples 253

50. Single phase loads connected to the load side of a phase converter shall not be connected in which of the following manners:

  a. To the manufactured phase   **b.** To the grounded phase
  c. To the neutral   **d.** To the high leg

ANSWERS

QUIZ 1—REVIEW

1. **D** Table 4, RNC - Articles 352 and 353
2. **B** Table 4, FMC - Article 348. Go to the *Over 2 Wires 40%* column
3. **B** If you are working with the same size conductors. (NOTE: (1) in Chapter 9 refers you to Appendix C to determine the appropriate conduit or tubing size. Table C8 lists the maximum number of conductors and fixture wires you can install in RMC. Since a 1-inch conduit is approved to hold 17 #10 THWN copper conductors, you have to go up to the next size for 18 conductors to 1 1/4 inches RMC.)
4. **B** FPN Number 2 in Table 1. (NOTE: Chapter 9 outlines the fact that if the ratio of the raceway (inside diameter) to the conductor or cable [outside diameter] is between 2.8 or 3.2, jamming can occur when pulling three conductors or cables into a raceway. To determine the ratio, first look up the conductor diameter in Table 5 and then the tubing diameter in Table 4. Since the approximate diameter of a 350kcmil THWN copper conductor is 0.817 inches and the internal diameter of a 2 1/2-inch EMT is 2.731 inches, the ratio of the EMT to the conductor is 2.731 /0.817 = 3.342 inches. This ratio falls outside the range of 2.8 to 3.2, which means jamming should not occur.)
5. **C** STEP 1: Calculate the total cross-sectional area of the 28 #12 THWN copper conductors using Table 5, Chapter 9. The cross-sectional area of each is .0133 square inches. This means the total conductor fill is:

    28 × .0133 square inches = .3724 square inches

    STEP 2: Select the conduit type and look up the total area 100% value in Table 4. Multiply this value by 60 percent, and if the value is greater than the total conductor fill value then the conduit size is acceptable. When you run the numbers using a 1-inch conduit, it has a 100% Total Area of .864 square inches.

    .864 × 60% = .5184 square inches

    Since .5184 . is greater than .3724 square inches., a 1-inch EMT nipple is acceptable.
6. **C** Note (9), Table 1, Chapter 9 determines that a multiconductor cable should be treated as a single conductor for calculating percentage conduit fill area. The area calculation of cables with elliptical cross-sections must be based on the major

diameter of the ellipse as a circle diameter. So first you have to calculate the cross-sectional area of the conductor:

Area = pi × 2 [(diameter)²/4] = 3.14159 × [(1.625)²/4] = 2.074 square inches.

Now if you look in the *1 Wire 53%* column in Table 4, you will see that 2 1/2 inches RMC is required for this application.

7. **A** STEP 1: Determine the cross-sectional area (in square inches) of each type of conductor in the run from Table 5.

$$\#6 \text{ Cu} = .0507 \text{ square inches}$$
$$\#8 \text{ Cu} = .0366 \text{ square inches}$$
$$\#10 \text{ Cu} = .0211 \text{ square inches}$$

STEP 2: Calculate the total square inches by multiplying the number of conductors by the cross-sectional value:

[2 × .0507] + [2 × .0366] + [4 × .0211] = .259 square inches

STEP 3: Refer to Table 4, which requires a 1-inch EMT. Finally, since the question used multiple wires, go to the *Over 2 Wires 40%* column to verify the EMT size.

8. **D** Annex B, Definition of Thermal Resistivity
9. **D** Table C.1(A)
10. **B** See Example D1(a) plus Example D1(b) Annex D

QUIZ 2

1.	B	14.	A	27.	B	39.	A
2.	C	15.	C	28.	B	40.	C
3.	A	16.	B	29.	B	41.	A
4.	B	17.	A	30.	A	42.	B
5.	B	18.	A	31.	C	43.	B
6.	C	19.	B	32.	D	44.	D
7.	B	20.	B	33.	B	45.	B
8.	C	21.	B	34.	D	46.	A
9.	B	22.	C	35.	A	47.	C
10.	B	23.	D	36.	C	48.	C
11.	A	24.	A	37.	D	49.	A
12.	D	25.	B	38.	D	50.	D
13.	A	26.	D				

QUIZ 3

1. C
2. B (NOTE: Arc-flash is not the same as Arc-Fault)
3. B
4. B
5. C
6. B
7. A
8. C
9. C
10. A
11. C
12. B
13. B
14. B
15. A
16. C
17. B
18. C
19. D (NOTE: the questions asks which do NOT have to be grounded)
20. B
21. B
22. C
23. C
24. A
25. A
26. A
27. C
28. D (NOTE: FCC goes under carpet SQUARES, not carpet)
29. A
30. A
31. C
32. B
33. D
34. B
35. A
36. D

> ▶ **Test** tip
> If you know for a fact that at least two of the answers are correct, in this case located outdoors or indoors, then you can assume that the correct answer is "D"-all of the above.

37. D
38. B
39. B
40. B
41. C
42. A
43. B
44. D
45. C
46. A
47. C
48. D
49. B
50. A

—NOTES—

Chapter 10
MATH CALCULATIONS AND BASIC ELECTRICAL THEORY

Electrician's Exam Study Guide

Remember back in school when you would complain about having to take algebra because you would never have to really use it out in the real world? Well, welcome to the wonderful world of mathematics. This is such serious stuff when it comes to electrical work that it even has its own laws. In order to pass the sections of the electrical exam that incorporate conversions and calculations, you will need to know which equations to use for each example provided. Bottom line: the answer may require you to determine volts, current or amperes, or voltage drops. Beyond the exam, once you get your license, your occupation will frequently require you to know the basics of Ohms law, fraction conversion, and other mathematical methods.

We will first have a quick review of basic math principles used in electrical calculations.

BASIC MATH SKILLS

How to Handle Fractions

Multiply fractions by putting the top numbers (the numerators) over the bottom numbers (the denominators) as shown in the example below:

$$\frac{1}{2} \times \frac{1}{8} = \frac{1 \times 3}{2 \times 8} = \frac{3}{16}$$

When you have to multiply a fraction by a whole number, put the whole number on top (numerator) over a 1 (denominator) as shown below:

$$\frac{1}{2} \times 5 = \frac{1 \times 5}{2 \times 1} = \frac{5}{2}$$ Now divide 2 into 5 to arrive at the lowest amount, which is 2 1/2.

When working with fractions, you need to reduce your answer to the lowest possible number. In the example below, 12 is the smallest common number. 12 goes into 12 one time and into 72 six times:

$$\frac{3}{8} \times \frac{4}{9} = \frac{3 \times 4}{8 \times 9} = \frac{12}{72} = \frac{1}{6}$$

Deciphering Decimals

A decimal is simply a fraction in which the denominator is not written out. The denominator value is represented as 1. Electrical instruments measure in decimals, not fractions, and manufacturer components list values in decimals, such as 1.5 amps. To convert a fraction to a decimal, divide the numerator (top number) by the denominator (bottom number) as shown below:

$$\frac{3}{4} = 3 \div 4 = 0.75$$

$$\frac{1}{2} = 1 \div 2 = 0.5$$

Practicing Percentages

To convert a percentage to a decimal, just move the decimal two places to the LEFT as shown below:
  50 percent is the same as 50.00

$$50.00 = 0.50$$

37 1/2 percent = 37.5 percent, which equals 0.375 (remember how you determined that 1/2 = 0.5 when you converted fractions to decimals).

Metric Conversion

There will be many times when you will need to perform metric conversions in the field and also on the licensing exam. The NEC® approves of two different types of conversions. The soft conversion technique is a direct conversion approach that uses number in their most finite representation, such as 2.988. The hard conversion method rounds off to the next whole value of a number, so that 2.988 is valued at 3. You will need to use these methods when you convert metric numbers to feet or inches in the field because few electricians measure the runs or cuts for cabling and conduit in meters. Additionally, you will encounter questions on the licensing exam that give examples in meters and answer options in feet, so you will need to know how to perform these conversions. Here are the basic conversions:

- 1 yard = 0.9144 meter
- 1 foot = 0.3048 meter
- 1 inch = 25.4 millimeters
- 1 meter = 1.09 yards or 39.36 inches or 3.28 feet

The examples above use the soft conversion method. Rounding to the nearest whole number gives the hard conversion value of 1 meter equals 3 feet.

Applying Math Principles

Throughout the NEC, requirements are listed in percentages. For example, section **[430.110(A)]** states that motor circuits rated 600 volts nominal or less must have a minimum ampere rating of 115 percent of the full-load current rating for the motor. Let's apply some math principles to a question based on this section.

A 480 volt motor with a full-load current of 25 amperes requires a disconnect rated at which of the following?

- **a.** 15 amperes
- **b.** 20 amperes
- **c.** 25 amperes
- **d.** 30 amperes

> ▶ **Test** tip
> Because the motor load is 25 amperes, you can eliminate a. (15) and b. (20) immediately.

ANSWER

25 amperes × 115% = 25 × 1.15 = 28.75.

Answer **d.** (30) is the closest to 115 percent.

Now we can look at the difference between elements such as an electrical conductor and an insulator, and how mathematic principles apply to calculating the fundamentals required to transfer electricity. Electricity flows easily through a **Conductor**. Copper is a good conductor of electricity and is commonly used in electrical wiring. The human body is also a good conductor of electricity. **Insulators** resist the flow of electricity. Insulating materials are used to coat copper conducting wires and are used to make electrical work gloves. Insulators help to protect humans from coming into contact with electricity flowing through conductors.

We have just stated that conductors allow electricity to flow through them while insulators resist the flow of electricity. Let's discuss some additional terms that are used to describe how electricity is transferred through conductors. These terms are *voltage, current,* and *resistance*. As an electrician, you already have experience working with these elements, but it's important to know the details of electrical energy. If, for example, you worked in the explosives field, it would not be enough to know that once you light the fuse, the dynamite goes boom.

Voltage (V) is the difference in the electrical potential between two specific points. Voltage can be thought of as electrical pressure—it is the force that causes electrical charge to move or flow through a system. If you wanted to get really specific, the definition of voltage is one Joule per Coulomb. A Joule is a unit of energy and a Coulomb is the amount of electrical charge. This means that a volt is the amount of energy in any given electrical charge.

Electric Current (I) is the rate of flow of electrical charges through a conductor. The charges can be positive or negative and are measured in amperes. An ampere is equal to one Coulomb per second.

Resistance (R) is measured in ohms and is, quite simply, the amount of resistance or opposition encountered by the flow of electrical charges through a conductor. The rate at which electricity flows through a circuit is affected by the resistance of the components and/or wires in the circuit. Increased resistance results in a decrease in the amount of current that will flow through the wire. Appliances, lights, and power tools can all be thought of as resistors.

Ohm's Law: Ohm's law is a mathematical formula used to describe the relationship between voltage (V), current (I), and resistance (R). Ohm's law can be written as follows:

$$I = \frac{V}{R}$$

Ohm's Law is described by a formula that illustrates the relationship between voltage, current, and resistance.

Algebra is the form of math that explores a mathematical statement by using letters or symbols to represent numbers. You will see Ohm's law represented in a variety of ways. Ohms or amps will often be illustrated using the algebraic statement:

$$I = \frac{E}{R}$$

For those of us who are not that fond of trying to remember what letter represents which part of the equation, perhaps the graphic below will be more helpful.

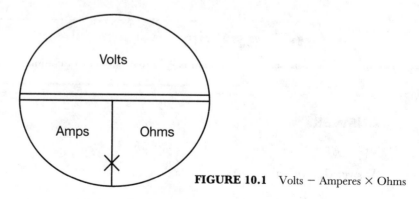

FIGURE 10.1 Volts − Amperes × Ohms

Math Calculations and Basic Electrical Theory

In this illustration, moving from right to left, you can visualize that Volts divided by Amperes equals Ohms and the same equation works from left to right: Volts divided by Ohms equals Amperes. Finally, if you only have the Amperes and the Ohms in a question, you can multiply them and determine the Volts.

Try this method on the test question below:

> If the circuit voltage of 120 volts is passed through a ground rod with the resistance of 25 ohms, how much current is passing through the grounding conductor?
>
> **a.** 0.21 A **b.** 0.48 A
>
> **c.** 4.8 A **d.** 5 A

The first step is to look at the possible answers and determine what part of Ohms law you are calculating. At a glance, you know that you are going to compute Amperes, because all of the answers are in Amps. Math is one of those unforgiving disciplines with infinite answers that does not allow for the "best guess" approach. When you know Ohm's law, figuring out the answer is as simple as:

$$\frac{120 \text{ Volts}}{25 \text{ Ohms}} = \mathbf{4.8} \text{ Amperes or answer "c".}$$

The same question could have looked something like this:

> If a circuit passes through a ground rod with the resistance of 25 ohms and draws 4.8 amperes of current through the grounding conductor, at what voltage is the circuit operating?

The answer would be 25 Ohms × 4.8 Amperes, which equals = 120 volts.

The first step to getting the correct answer is to know the rules surrounding this appliance. Section **[422.11(E)(3)]** states that if the overcurrent protection rating isn't marked on the appliance, and the appliance is rated over 13.3 amperes, then the overcurrent protection (in this case the circuit breaker) can't exceed 150 percent of the appliance rating. It then goes on to say, and this is the important part as it relates to your exam answer, that if 150 percent of the rating does not correspond with standard overcurrent device ratings, then the next higher size should be used.

Let's specifically apply the code requirements to the calculation process. Assume that one of the test questions you have on your exam is based on Section **[422.11(E)]**, which covers single non-motorized appliances. The question is:

> If a branch circuit supplies a quick recovery electrical water heater, which is rated at 4500 watts at 240 volts, which of the following circuit breaker sizes is required to protect the appliance:
>
> **a.** 15 amp **b.** 18.75 amp
>
> **c.** 20 amp **d.** 30 amp

Not only do you have to know which code rules to apply, but you also need to know how to do the math involved to establish the correct answer. First we have to use Ohm's law to calculate the amperes. Take the 4500 watts and divide them by the 240 volts.

4500 ÷ 240 = 18.75 but you are not done yet. If you stop here, you might pick answer B, but that answer does not take into consideration the code requirements for this non-motorized appliance. Therefore, you now have to multiply that value by the 150 percent required by the code, so you have to convert the percentage to a whole number.

$$1{\underset{\smile}{50.00}} = 1.50 \quad \text{Now you can multiply } 18.75 \times 1.50 = 28.1$$

This size, 28.1, is not a standard breaker rating, so you are allowed to use the next size up, which would be 30 amp, or answer **d**.

Power is the rate at which energy is expended and is measured in watts or kilowatts. One watt is equal to one Joule per second. In order to calculate power watts, you have to know the voltage, current, and the power factor.

Power Factor (pf) only occurs in an alternating current circuit and can range from 0 to 1. The induction and capacity of the circuit can vary the voltage sine wave and the current sine wave so that they don't peak or reach zero at precisely the same time. When this occurs, the power factor drops below one. The power factor of a circuit will be lower depending on the degree of misalignment between the current and voltage. Because electric motors have inductance, they usually have a power factor of less than 1. An incandescent light bulb has resistance, so the voltage and the current are aligned and the power factor will be 1.

The graphic below illustrates how to calculate Watts of power:

$$\text{Power/watts} = \boxed{\text{Volts}} \times \boxed{\text{Amps}} \times \boxed{\text{power factor (pf)}}$$

There are three wires in a 3-phase circuit, instead of just the two wires found in a single-phase circuit. One conductor in a 3-phase load has a current that is shifted 120 degrees from the current running in each of the other two wires. This means that even if the power, voltage and power factor are the same, less current is actually flowing through a 3-phase circuit than through a single-phase circuit with the same watts. When dealing with 3-phase calculations, you need to remember that the square root of 3 is 1.73.

> ### ▶ **Math** tip
>
> **Square Roots**
>
> The square root of a number, n, shown below is the number that gives n when multiplied by itself.
>
> n^2 and $\sqrt{n}$ are both ways to represent a number squared. Don't confuse this mathematical process with addition.
>
> You know that $1\frac{1}{2}$ is the same as 1.5
>
> $$\begin{array}{r} 1.5 \\ +1.5 \\ \hline 3.0 \end{array} \quad \text{BUT } 1.5 \times 1.5 \text{ only equals } 2.25$$
>
> 1.73 (or "n") × 1.73 ("n" again) = 3

This number is not usually supplied in the test question because the example assumes that you know the square root of 3 = 1.73. If a test question is based on a 3-phase circuit example, then you would use the same Watt graphic above, but include 1.73 in the equation as follows:

$$\text{Power/watts} = \boxed{1.73} \times \boxed{\text{Volts}} \times \boxed{\text{Amps}} \times \boxed{\text{power factor (pf)}}$$

CALCULATING TRANSFORMERS

Transformers change voltage and consist of two separate coils of conductor wound around a laminated steel coil. Alternating current goes through one of the coils of conductor, and the current generates a magnetic field around the coil. Since the current is moving back and forth, the magnetic field is always in motion. The second coil is usually wound right over the first so it is within the magnetic field created by the first coil. The result is that the magnetic field moves and induces current to flow through the second coil of the conductor. The voltage in the first coil is based on the amount of turns around the steel coil and is directly proportional to the number of turns of the second coil.

The first coil is called the "**primary**" winding and the second coil is called the "**secondary**" winding. The **turn ratio** is the comparison in the number of turns each coil makes. For example, if the primary has twice as many turns as the secondary, then the turn ratio is 2. The equation looks like this:

$$\frac{\text{Voltage (primary)}}{\text{Voltage (secondary)}} = \frac{\text{Number of turns (primary winding)}}{\text{Number of turns (secondary winding)}} = \text{TURN RATIO}$$

Now we can take this principle one step forward and establish a relationship between the volts and amperes in the primary with those in the secondary. When installing a transformer and the wiring and overcurrent protection for the transformer, you need to know that the volts multiplied by the amperes of the primary is equal to the volts multiplied by the amperes of the secondary.

$$\text{Primary Volts} \times \text{Primary Amperes} = \text{Secondary Volts} \times \text{Secondary Amperes}$$

This equation is illustrated in the sample question below:

If the secondary coil of a transformer produces 100 amperes at 240 volts, and the primary is energized at 480 volts, how many amperes does the transformer primary coil produce?

- **a.** 100 amperes
- **b.** 200 amperes
- **c.** 240 amperes
- **d.** 50 amperes

The short cut to the correct answer is to start with the information provided on the secondary:

Secondary Volts of 240 × Secondary Amperes of 100 = **24,000**

Since you know there are 480 volts in the primary, and you also know that the total of the volts × the amperes in the secondary must equal the total of the volts (480) × the amperes of the primary, you can simply DIVIDE the total you got for the secondary (24,000) by the 480 volts you know you have in the primary.

$$\frac{24000 \text{ (total of the volts} \times \text{amperes in the secondary)}}{480 \text{ (volts in the primary)}} = 50$$

This then proves out the equation we demonstrated above:

Secondary Volts (**240**) × Secondary Amperes (**100**) = Primary Volts (**480**) × Primary Amperes (**50**)

The first thing to remember is how to determine the turn ratio. You need to know the turn ratio in order to figure out how to calculate the amperes.

Here is another example of transformer power theory:

If the primary coil of a transformer produces 480 volts, and the secondary winding produces 240 volts, what is the turn ratio of the transformer?

a. .2   **b.** 2

c. 240 amperes   **d.** 50 amperes

$$\frac{\text{Voltage 480 (primary)}}{\text{Voltage 240 (secondary)}} = 2 \text{ which is the TURN RATIO}$$

This lets you know that you will have to divide the secondary amperes of 100 by 2, leaving you 50 amperes in the primary.

Now that you have taken the time to practice applying your math skills with principles of basic electrical theory, it's time for another test. Get out your calculator and code book, because this is a simulated timed test.

TIMED MATH TEST

TIME ALLOWED: 60 MINUTES

1. The load that can be used for the service calculation for a dwelling unit with a range rated at 13 KW is which of the following:
   - **a.** 13 kw
   - **b.** 8.4 kw
   - **c.** 8 kw
   - **d.** 5.5 kw

2. A motor that draws 4.5 kilowatts from the power line and delivers 5 horsepower has an efficiency equal to which of the following:
   - **a.** 90.5 percent
   - **b.** 83 percent
   - **c.** 75 percent
   - **d.** 45 percent

3. Based on the figure below, the voltage across the 5 amp load after the neutral is opened, if both loads have resistance and the power factor is 1.0, would be which of the following:

   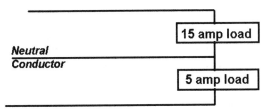

   - **a.** 240 volts
   - **b.** 180 volts
   - **c.** 150 volts
   - **d.** 100 volts

4. If the raceway is 12 feet long, the allowable size for a rigid steel conduit with four 4/0 XHHW conductors and one #3 bare equipment grounding conductor would be which of the following:
   - **a.** 1 inch
   - **b.** 1 1/2 inches
   - **c.** 2 inches
   - **d.** 2 1/2 inches

5. If an air conditioner motor compressor unit consumes 8500 volt amperes of power at a line voltage of 240 volts, then the minimum total circuit ampacity shall be which of the following:
   - **a.** 35.4 amps
   - **b.** 44.3 amps
   - **c.** 70.2 amps
   - **d.** 85 amps

6. A four-story office building has a required load of 60000 volt amperes and single-phase 240 volt feeder circuits supplying the general lighting load, which means that the minimum required current in the feed circuit ungrounded conductors is which of the following:
   - **a.** 240 amps
   - **b.** 250 amps
   - **c.** 270 amps
   - **d.** 480 amps

7. A 1500 square foot single-family, one-bath dwelling has a 12 kW electric range and a 5.5 kW, 240 volt dryer. The general lighting load is which of the following:

   a. 1500 VA
   b. 2500 Va
   c. 4000 VA
   d. 4500 VA

8. This same 1500 square foot single-family, one-bath dwelling with a 12 kW electric range and a 5.5 kW, 240 volt dryer has a general lighting load in amperes of which of following:

   a. 20 amps
   b. 33.5 amps
   c. 37.5 amps
   d. 40 amps

9. The minimum number of branch circuits required for 12 kW range in this same 1500 square foot single-family, one-bath dwelling would be which of the following:

   a. Two 2-wire 20 amp circuits
   b. Three 15 amp 2-wire circuits
   c. One 2-wire 20 amp circuit
   d. Two 2-wire 15 amp circuits

10. Provided that the calculation for the neutral for the feeder and service is 14550VA, then the calculated load for neutral for this same 1500 square foot single-family, one-bath dwelling would be which of the following:

    a. 37.5 amps
    b. 40.4 amps
    c. 60.6 amps
    d. 77.5 amps

11. A church building has the outside dimensions of 100 feet x 200 feet, which means that the minimum lighting load required for the general lighting only would be which of the following:

    a. 2000 volt amperes
    b. 10000 volt amperes
    c. 12000 volt amperes
    d. 20000 volt amperes

12. If there are three resistors in a series circuit, labeled R1, R2, and R3, with a total power used of 225 watts, and the individual power used by R2 is 75 watts and by R3 is 100 watts, then the power used by resistor R1 is which of the following:

    a. 100 watts
    b. 75 watts
    c. 50 watts
    d. 25 watts

13. In a 100 volt circuit with a resistance of 25 ohms, the current flowing through the circuit is which of the following:

    a. 1.25 amps
    b. 2.5 amps
    c. 4 amps
    d. 5 amps

Math Calculations and Basic Electrical Theory

14. Three parallel circuits with a voltage of 150 volts each produce a total current of which of the following:
    a. 50 volts
    b. 150 volts
    c. 225 volts
    d. 450 volts

15. If a parallel circuit has thee circuits, C1, C2, and C3, with a total current of 600 volts, then the individual current in each circuit is equal to which of the following:
    a. 600 divided by the resistance of each circuit
    b. 200 volts
    c. 180 volts
    d. It is impossible to calculate without knowing the value of each circuit

16. A 60-cycle AC circuit of 120 volts and 12 amperes has a true power value of which of the following:
    a. 1220 volt-amperes
    b. 1220 watts
    c. 720 volt-ampers
    d. 720 watts

17. If a 480 volt motor has a full-load current of 34 amperes, then the standard disconnecting means must be which of the following:
    a. 66 amps
    b. 50 amps
    c. 39.1 amps
    d. 40 amps

18. A 32 watt metal halide fixture that uses .578 percent of its total wattage has a reciprocal wattage of which of the following:
    a. 31.42
    b. 1.73
    c. .422
    d. 5%

19. An electric motor that runs at 3/4 of its full-load potential of 200 amperes is operating at which of the following:
    a. 150 amps
    b. 75 amps
    c. 25 amps
    d. 15 amps

20. Three 9-feet long sections of 1 1/2 inch rigid metal conduit can be installed in which of the following:
    a. A trench that is 42 inches deep and concrete encased
    b. A trench that is 8.2296 meters long
    c. A trench that is 3 feet × 3 feet
    d. A trench that is 2.7432 meters long

21. The total connected load for three volt electric water heaters, where the first water heater (WH1) is 1200 watts, 120 volts, the second water heater (WH2) is 1800 watts at 240 volts and the last water heater (WH3) is 1600 watts at 120 volts is equal to which of the following:

    a. 9200 watts   **b.** 7200 watts

    c. 5800 watts   **d.** 4600 watts

22. A 10 amp, 240 volt 3-phase line-to-line electric water heater with a power factor of 1.0 produces which of the following:

    a. 2400 watts of heat   **b.** 4152 watts of heat

    c. 4315 watts of heat   **d.** 4800 watts of heat

23. An electric coffee maker rated at 1200 watts and an electric toaster rated at 1440 watts are both plugged into a 120 volt outlet and will pull a combined amperage equal to which of the following:

    a. 2 amps   **b.** 22 amps

    c. 24 amps   **d.** 48 amps

24. If a 12 ohms resistor and a 6 ohms resistor and a 4 ohms resistor are connected in parallel, then the total resistance of the circuit must be which of the following:

    a. 22 ohms   **b.** 11 ohms

    c. 8.2 ohms   **d.** 2 ohms

25. A circuit voltage of 120 volts passing through a ground rod with the resistance of 20 ohms has a total current passing through the grounding conductor equal to which of the following:

    a. 0.20 amps   **b.** 0.60 amps

    c. 4 amps   **d.** 6 amps

ANSWERS

TIMED MATH TEST

1. **B** 8kw + 5 percent of 8kw = 8.4kw
2. **B** 5 × 746 watts (4500 watts = 83
3. **B** R = 15 a load so 120 ÷ 15 = 8 ohms; R = 5 a load so 120 ÷ 5 = 24; I for neutral open would then be 240 ÷ (8 + 24) = 7.5 amps; E across 5 a load = 7.5 amps × 24 = 180 volts
4. **C** Reference Section [344.22] and Chapter 9, tables 1, 4, 5 and 8. The area of four 4/0 is 4 × .3197 = 1.2788; the area of one #3 is .058; 1.278 + .053 = 1.3318 square inches. A 2-inch rigid conduit allows for 40 percent of the area of 1.363 square inches.
5. **B** Reference Table [440.32]. I = E ÷ R, so take 8500 ÷ 240 = 35.4; from Table [440.32] take 125 percent of 35.4 = 44.3 amps
6. **B** Reference Chapter 9, Annex D. I = E ÷ R, so take 60000 ÷ 240 = 250 amps
7. **D** Reference Section [220.40] 1500 ft^3 at 3 VA per foot^3 = 4500 VA
8. **C** 4500 VA ÷ 120 volts = 37.5 amps
9. **A** Reference Section [210.1(C)(1)]
10. **C** 145500 VA ÷ 240 volts = 60.6 amps
11. **D** 100 inches × 200 inches = 20,000 square feet. Table [220.12] lists 1 VA per foot required for churches, so 1 × 20000 = 20000 VA
12. **C** Take the total power of 225 − R2 value of 75 = 150; 150 − R3 value of 75 = 50 watts
13. **C** Ohm's law I = E ÷ R, so 100 ÷ 25 = 4 amps
14. **D** Total current is the sum of all of the circuits. 150 + 150 + 150 = 450
15. **B** In a parallel circuit the voltage is the same across each resistance load; therefore, you can divide the total load of 600 by 3 circuits and know that each circuit has a value of 200 (600 ÷ 3 = 200).
16. **D** 120 volts × 12 amps = 1440 volt-amperes (which is the apparent power) 1440 × .5 = 720 watts
17. **D** Reference Section [430.110(A)], which states that the disconnecting means for motors rated 600 volts or less must be at least 115 percent of the full-load current rating of the motor. Therefore, you would take 115 percent (or 1.15) × 34 amps = 39.1 amps. However, the question specifically asks for the standard disconnect size, and since 39.1 is not an industry standard, the answer is 40 amps.
18. **B** This question has nothing to do with the total wattage of 32 watts. The only thing the question is asking for is the reciprocal value of .578, which is calculated by dividing 1 by .578.
19. **A** 3/4 is 75 percent or .75, so 200 amps × .75 = 150 amps.

20. **B** It does not matter that the conduit is 1½ inches. The answer is the conversion of yards to meters. 1 yard = 0.9144 meters. A 9-feet section of conduit is 3 yards long, so 3 × 0.9144 = 2.7432 meters; however, there are three sections of the conduit, so you have to multiply 2.7432 × 3 = 8.2296 meters

21. **D** The voltage has nothing to do with calculating the answer. The answer is simply the sum of all wattages for all of the water heaters, so 1200 + 1800 + 1600 = 4600 watts.

22. **B** Since this is a 3-phase system and you need to find the power, you take 1.73 (for 3-phase) × 240 volts × 10 amps × 1.0 (power factor) = 4152 watts.

23. **B** Amperes = Power ÷ Volts. 1440 watts ÷ 120 volts = 12 amps and 1220 ÷ 120 = 10 amps. Add 12 amps + 10 amps for a total load of 22 amps.

24. **D** Resistance total $(R_T) = \dfrac{1}{R_T} \dfrac{1}{12} + \dfrac{1}{6} + \dfrac{1}{4} = \dfrac{6}{12} \quad \dfrac{12}{6} = 2$ ohms

25. **D** Volts ÷ Amps or 120 ÷ 20 = 6 amps

Chapter 11
REVIEW AND APPLYING PRINCIPLES

Electrician's Exam Study Guide

We'll take some time now to practice applying the principles discussed in the NEC® to scenarios you might encounter in the field. These exercises are meant to show you the connection between various Code requirements and how you may have to integrate them into your work once you have your license.

For example, let's say you are working on a three-wire 120/240 volt system with 3-phase conductors and a grounded conductor, and the hot wires are connected to different phases. By definition, what kind of circuit are you working with? According to Article 100, this is a multi-wire branch circuit. Perhaps you are about to estimate a commercial remodeling job where there is an existing exposed 800 volt conductor located in the corner of an open storage area. What do you have to include in your pricing in order to meet current codes? Unless there is a qualified person stationed to observe the equipment at all times (highly unlikely but a great job if you can get it), the area will need to have walls installed with a lockable door that can be unlocked from the inside (Section **[100.76(B)]**), and it must be labeled DANGER – HIGH VOLTAGE – KEEP OUT. This requirement is described in Section **[110.34(C)]** under Locked Rooms or Enclosures.

BASIC PRINCIPLES

You want to be able to estimate the cost of a job to maximize your profit and minimize your material waste. So let's apply the NEC® code requirements to a basic box installation. Assume that we are working with a metal box that is 4 inches × 4 inches × 1 1/2, and we are using 12 AWG wire. Our job is to determine how many conductors can run into the box based on several conditions. First of all, our installation has a raceway connector, one equipment bonding jumper that terminates inside the box, a black unbroken conductor that passes through the box, and two white conductors that originate outside of the box and are spliced inside with a wire connector. Additionally, we are going to install a box on each side of a fire-resistant rated wall. The sketch below illustrates one of the two typical boxes in this layout.

Table **[314.16(A)]** indicates that by using 12 AWG wire, we can have up to nine conductors in this box. Section **[314.16(B)(1)]** determines how the conductors are counted. The raceway connector is not counted. The conductor passing through the box counts as one. The conductors that originate outside of the box and are terminated with a wire connector count as two conductors, even though they are connected. A conductor that does not leave the box, such as our bonding jumper, is not counted. The conductors that originate outside of the box and are spliced inside of it count as a total of two conductors. This means that

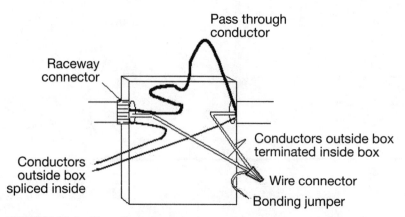

FIGURE 11.1 You will need to know how to count conductors inside and outside a box.

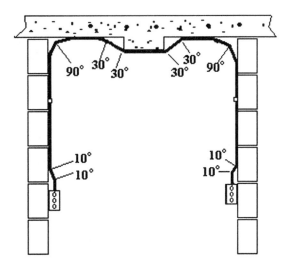

FIGURE 11.2 The total number of bends can not exceed 360 degrees.

our example has a total of five conductors, which is well below the maximum of twelve allowable conductors. Therefore, each box design conforms with the NEC® requirements. Finally, we need to take into consideration that each of these boxes is going to be installed on opposite sides of a fire-resistant wall. This means that there must be 24 inches of horizontal clearance between each box location in order to conform with Section **[300.21]** (**FPN**).

Next, let's consider how to calculate the total number of allowable bends in a raceway. Our raceway uses EMT conduit and runs from a surface mounted junction box, up the side of a block wall, across the top of a concrete ceiling, under a concrete support beam, and across to a surface mounted box on the opposing block wall. The sketch above illustrates the various degrees of bends in the run.

Section **[358.26]** determines that the total bend maximum for EMT is 360 degrees. This is the equivalent of four quarter bends (which would total 360 degrees) maximum between pull points such as conduit and boxes. All of the bends must be counted, even those coming out of the junction box. This means that the two angles needed to come out of the surface mounted box and over to the wall are each 10-degree bends and count as a total of 20 degrees just to angle back to the wall. If you total the bends in our diagram, we come up with 340 degrees which conforms to the maximum 360 degrees allowed by the NEC. Finally, the raceway installation must be complete between an outlet, junction, or splicing point prior to the installation of the conductors, per Section **[300.18(A)]**.

ADVANCED PRINCIPLES

Now we can move up to some more specific calculations. We can start with a branch circuit installation. Our branch circuit is supplied by a 208/120 volt, 4-wire, 3-phase, wye connected system. What size branch-circuit overcurrent protection device and conductor (THHN) is required if we are using a 19 kVA of non-linear loads (75 degrees C terminals)?

Step 1: First we need to size the overcurrent protection device in accordance with Sections **[210-20(A)]** and **[384-16(D)]**. In order to accomplish this, we have to convert the nonlinear load from kVA to amperes. Keep in mind that our example deals with a 3-phase system, so we have to multiply our amperes by 1.732.

$$\text{Amperes} = \text{VA}/(\text{Volts} \times 1.732)$$

$$\text{Amperes} = 19{,}000/(208 \text{ volts} \times 1.732)$$

$$\text{Amperes} = 52.74 \text{ amperes, which rounds up to 53 amperes}$$

The branch-circuit overcurrent protection device cannot be sized less than 125 percent of these amperes.

$$53 \text{ amperes} \times 125\% = 66 \text{ amperes}$$

According to Section **[240-6(A)]**, we have to choose a minimum 70 amperes overcurrent protection device. Now we need to find the conductor that complies with Sections **[110-14(C)]** and with Section **[210-19(A)]** which also requires the branch-circuit conductor to be sized no less than 125 percent of the continuous load.

Step 2: Since we came up with 53 amperes during our initial conversion in Step 1, we multiply that by 125% required in Section **[210-19(A)]** and we get 66 amperes. We have to use the conductor according to the 75 degrees C temperature rating of the equipment terminals. No. 6 THHN has a rating of 65 amperes at 75 degrees C, so we *cannot* use it; therefore, we have to go to the next size of #4 THHN, which has a rating of 85 amperes at 75 degrees C.

Step 3: Our #4 THHN conductor has to be protected against overcurrent in accordance with Section **[240-3]**. First, we have to make sure that the No. 4 THHN will be properly protected against overcurrent by the 70 ampere overcurrent protection device. Since we have more than three current-carrying conductors in the same raceway, we have to adjust the No. 4 THHN conductors ampacity based on the listing in the 90 degrees C column of Table **[310-16]**.

$$\text{Corrected Ampacity No. 4 THHN} = \text{Ampacity} \times \text{Note 8(A) Adjustment Factor}$$

$$\text{Corrected Ampacity No. 4 THHN} = 95 \text{ amperes} \times 80 \text{ percent Corrected Ampacity}$$

$$\text{No. 4 THHN} = 76 \text{ amperes}$$

Now the #4 THHN, which is rated at 76 amperes after ampacity correction, will be properly protected by using a 70 ampere overcurrent protection device and complies with the general requirements of Section **[240-3]**.

Test questions on the exam that address this same concept could include some of the following:

In a 208/120 volt, 4-wire, 3-phase, wye connected branch circuit system, which uses a 19 kVA of nonlinear loads (75 degrees C terminals), the total amperes used are which of the following:

a. 30   **b.** 53
c. 54   **d.** 60

In a 208/120 volt, 4-wire, 3-phase, wye connected branch circuit system, which uses a 19 kVA of nonlinear loads (75 degrees C terminals), the size branch-circuit overcurrent protection device to be used is which of the following:

a. 54 amperes   **b.** 60 amperes
c. 70 amperes   **d.** 100 amperes

See how understanding the process involved with determining device sizes, as well as the Code requirements, can help you come up with the most accurate, correct answer? Let's try another example using feeder continuous loads. Ready?

Determine the required size for the feeder overcurrent protection device and conductor (THHN) for a 184 ampere continuous load on a panelboard (75 degrees C terminals) that supplies nonlinear loads. The feeder is supplied by a 4-wire, 3-phase, wye connected system.

First, size the overcurrent protection device based on the requirements in Sections **[215-3]** and **[384-16(D)]**. The feeder overcurrent protection device must be sized *not less than* 125 percent of the 184 amperes.

$$184 \text{ amperes} \times 125\% = 230 \text{ amperes}$$

According to Section **[240-6(A)]** we have to use a *minimum* 250 ampere overcurrent protection device.

Determine the proper conductor that complies with Sections **[110-14(C)]** and **[215-2]**, remembering that Section **[215-2]** requires that the feeder conductor cannot be sized *less than 125 percent of the continuous load*.

$$184 \text{ amperes} \times 125\% = 230 \text{ amperes}$$

Using the conductor according to the 75 degrees C temperature rating of the panelboards terminals, we can determine that No. 4/0 THHN has a rating of 230 amperes at 75 degrees C. We have to use the 250 amperes rating.

Finally, we need to know that a No. 4/0 conductor has to be protected against overcurrent in accordance with Section **[240-3]**. Since we have more than three current-carrying conductors in the same raceway, we have to adjust the No. 4/0 THHN conductors ampacity based on the requirements listed in the 90 degrees C column of Table **[310-16]**.

$$\text{Corrected Ampacity No. 4/0 THHN} = \text{Ampacity} \times \text{Note 8(A) Adjustment Factor}$$

$$\text{Corrected Ampacity No. 4/0 THHN} = 260 \text{ amperes} \times 80\%$$

$$\text{Corrected Ampacity No. 4/0 THHN} = 208 \text{ amperes}$$

Unfortunately, the No. 4/0 THHN, which is rated 208 amperes after we calculated ampacity correction, is not considered protected by a 250 ampere overcurrent protection device. Why? Because of the "next size up rule" in Section **[240-3(B)]**, which only permits a 225 ampere protection device on the 208 ampere conductor—see Section **[240-6(A)]**. This means we have to bump the conductor size up to 250 kcmil to comply with the overcurrent protection rules found in Section **[240-3]**.

Now we are going to cover a quick review of current flows and grounding requirements. Electricity is a force that wants to run rampant. Given multiple conductive paths, current will not simply seek out the path of least resistance. According to Kirchoff's current law, when in parallel paths, current will divide and run though each available path. Although more current will flow through the path of least resistance, it will nonetheless flow through all available paths.

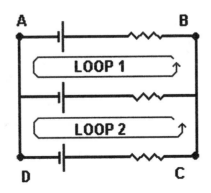

FIGURE 11.3 Kirchoff's law.

When we ground current flow, Section **[250]** tells us that it is critical to remember the only way to safely protect against a ground fault is to bond electrical equipment to an effective ground-fault current path. You will often see exam questions that address grounding equipment to the earth. The following is a typical example:

> Metal parts using 120 volts must be grounded to a suitable grounding electrode so that in the event of a ground fault, current will be shunted into the earth and protect unqualified persons from electric shock.
>
> **a.** True **b.** False

There are several elements to consider in this question. First, it doesn't matter if the person in the example is qualified or unqualified, so don't let that part throw you off. The bottom line is whether grounding to the earth will protect anyone against electric shock. If a person touches an energized electric pole that is only grounded to the earth, they will receive a current flow shock of 90 to 120 mA. Additionally, because of the earth's high level of resistance, the requirements in Section **[250]** state that the earth does not provide an effective ground-fault current path. Given this information, the answer to the question is B) False.

You may sometimes need to determine voltage drop, which requires you to know the basics of what voltage drop is and what factors affect it. While the NEC® does not require you to size conductors to accommodate voltage drop, Fine Print Notes [FPN] in Sections **[210.19(A)]**, **[215.2(A)(4)]**, **[230.31(C)]** and **[310.15(A)(1)]** recommend that you adjust for voltage drop when sizing conductors. For best performance purposes, the NEC® also recommends that the maximum combined voltage drop for both the feeder and branch circuit should not exceed 5 percent, and the maximum on the feeder or branch circuit should not exceed 3 percent.

You may be wondering, if these are only recommendations and not code requirements, then why spend and time on the subject? Well, your goal is to get your electrical license, and once you have that license you will need to know how to calculate things like voltage drop in the field. Some additional points are:

- Consider system efficiency as well as performance. If a circuit supports a large portion of a power load, then you will find that a larger conductor will pay for itself many times over in energy savings for your customer. Furthermore, lighting loads perform best when the voltage drop is minimal, which means you can effectively achieve the lighting performance of a higher-watt system simply by running larger wires.

- Undervoltage for inductive loads can cause overheating and inefficiency, and can drastically reduce the life span of equipment and wiring. When a conductor resistance could cause the voltage to drop below an acceptable point, increase the conductor size.

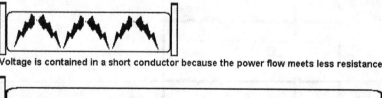

FIGURE 11.4   Voltage drop.

Review and Applying Principles

The voltage drop of a circuit is in direct proportion to the resistance of the conductor and the size of the current. If you increase the length of a conductor, you increase its resistance, which means you increase its voltage drop.

If you increase the current, you increase the conductor voltage drop. This is why long conductor runs often produce voltage drops that exceed NEC® recommendations.

Try this practice problem:

> The minimum recommended operating voltage for a 115V load connected to a 120V source is which of the following:
>
> **a.** 120 volts **b.** 115 volts
>
> **c.** 114 volts **d.** 116 volts
>
> The answer is **c.** 114V.

The maximum conductor voltage-drop recommended for both the feeder and branch circuit is 5 percent of the voltage source (120V).

The total conductor voltage drop (feeder and branch circuit) should not exceed 120V × 0.05 = 6V.

To calculate the operating voltage at the load, subtract the conductor voltage drop from the voltage source: 120V − 6V = 114V.

You can determine conductor voltage drop using either the Ohm's law method of $(I \times R)$ for single-phase calculations or the formula method. Keep the following principles in mind:

> For conductors 1/0 AWG and smaller, the difference in resistance between DC and AC circuits is negligible and not a factor. You can also disregard the small difference in resistance between stranded and solid wires.
>
> VD = Voltage Drop
>
> I = The load in amperes at 100% for motors or continuous loads.
>
> R = Conductor Resistance, refer to NEC® Chapter 9, Table 8 for DC or Chapter 9, Table 9 for AC.

Here is another sample problem:

> The voltage drop of two 12 AWG THHN conductors that supply a 16A, 120V load located 100 feet from a single-phase power supply is equal to which of the following:
>
> **a.** 3.2 volts **b.** 6.4 volts
>
> **c.** 9.6 volts **d.** 12.8 volts
>
> The answer is **b.** 6.4V.

Let's solve the problem using Ohm's law:

$VD = I \times R$

$I = 16A$

R = 2V per 1,000 feet, Chapter 9, Table 9: (2V/1,000 feet) × 200 feet = 0.4V

$VD = 16A \times 0.4V = 6.4V$

Formula Method

This method is a bit more complex, but you can use it for three-phase calculations that you can't do with the Ohms law method. Key elements of the formula method include:

- Single-phase $VD = 2 \times K \times I \times D/CM$.
- Three-phase $VD = 1.732 \times K \times I \times D/CM$. The difference between this and the single phase formula is that you replace the 2 with 1.732.
- K = Direct-current constant. K represents the DC resistance for a 1,000-circular mils conductor that is 1,000 feet long at an operating temperature of 75 degrees C. K is 12.9 ohms for copper and 21.2 ohms for aluminum.
- Q = Alternating-current adjustment factor: For AC circuits with conductors 2/0 AWG and larger, you must adjust the DC resistance constant K for the effects of self-induction (eddy currents). Calculate the "Q" adjustment factor by dividing the AC ohms-to-neutral impedance listed in Chapter 9, Table 9 by the DC resistance listed in Chapter 9, Table 8.
- I = Amperes: The load in amperes at 100% (not at 125% for motors or continuous loads).
- D = Distance: The distance of the load from the power supply. When calculating conductor distance, use the length of the conductor—not the distance between the equipment connected by the conductor. To arrive at this length, add distance along the raceway route to the amount of wire sticking out at each end. An approximation is good enough. Where we specify distances here, we are referring to the conductor length.
- CM = Circular-Mils: The circular mils of the circuit conductor as listed in NEC® Chapter 9, Table 8.

If we have a 3Ø, 36 kVA load rated 208V is wired to the panelboard with 80-feet lengths of 1 AWG THHN aluminum, what is the approximate voltage drop of the feeder circuit conductors?

How did we do that? Applying the three-phase formula, where:

K = 21.2 ohms, aluminum

I = 100A (36,000/(208 × 1.732))

D = 80 feet

CM = 83,690 (Chapter 9, Table 8)

VD (voltage drop) = **1.732** (for 3 phase) **× 21.2** (direct current constant) × **100A × 80 ft** √ **83,690** (circular mils) = **3.51 VD**

Using basic algebra, you can apply the formula method to find one of the other variables if you already know the voltage drop.

Here is an example. Suppose you have a 3Ø, 15 kVA load rated 480V and 390 feet of conductor. What size conductor will prevent the voltage drop from exceeding three percent?

K = 12.9 ohms, copper

I = 18A (15,000/(480 × 1.732))

D = 390 feet

VD = 480V × 0.03 = 14.4V

CM = 1.732 × 12.9 × 18 × 390/14.4V = 10,892, 8 AWG, Chapter 9, Table 8

Let's figure out what size conductor you need to reduce the voltage drop to the desired level. Simply rearrange the formula. For three-phase, it would look like this: $CM\,(3\emptyset) = 1.732 \times K \times I \times D/VD$. Of course, for single-phase calculations, you would use 2 instead of 1.732.

You could also rearrange the formula to solve a problem like this one: What is the maximum length of 6 AWG THHN you can use to wire a 480V, 3Ø, 37.5 kVA transformer to a panelboard so voltage drop does not exceed three percent (**see NEC® Figure 8-19**)?

$D\,(3\emptyset) = CM \times VD/1.732 \times K \times I$

$CM = 26{,}240$, 6 AWG Chapter 9, Table 8

$VD = 480V \times 0.03 = 14.4V$

$K = 12.9$ ohms, copper

$I = 45A\ (37{,}500/480 \times 1.732)$

$D = 26{,}240 \times 14.4/1.732 \times 12.9 \times 45 = 376$ ft

Sometimes, the only method of limiting voltage drop is to limit the load. Again, we can rearrange the basic formula algebraically: $I = CM \times VD/1.732 \times K \times D$. Suppose an installation contains 1 AWG THHN conductors, 300 feet long in an aluminum raceway, to a 3Ø, 460/230V power source. What is the maximum load the conductors can carry without exceeding the NEC® recommendation for voltage drop (**see Figure 8-21**)? Let's walk it through:

$I = CM \times VD/1.732 \times K \times D$

$CM = 83{,}690$ (1 AWG), Chapter 9, Table 8

$VD = 460V \times 0.03 = 13.8V$

$K = 12.9$ ohms, copper

$D = 300$ ft

$I = 83{,}690 \times 13.8/1.732 \times 12.9 \times 300 = 172A$

Note: The maximum load permitted on 1 AWG THHN at 75 degrees C is 130A **[110.14(C)] and Table [310.16]**.

This next application of code standards requires you to find the voltage across the 5 ampere load in the example below, after the neutral is opened.

First find the resistance of each load:

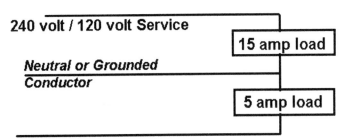

FIGURE 11.5 Voltage across resistance load.

15 ampere load:

Volts = 120

$R = E/I$

R = 120 volts/15 amperes so R = 8 ohms

5 ampere load:

R = 120 volts/5 amperes so R = 24 ohms

When neutral is open, the two resistors are in series and the Total Resistance is 8 + 24, or 32 ohms.

The voltage is 240 volts across the two resistors in series.

To find the current after the neutral opens:

I = 240 volts/32 ohms

I = 7.5 amperes

The voltage across the 5 ampere load that has a resistance of 24 ohms is:

$E = I \times R$

E = 7.5 ampere × 24 ohms

E = 180 volts

So the voltage across the 5 amp load is 180 volts.

Now let's look at how to determine the size of rigid metal conduit that would be permitted for 9 conductors, as listed below:

- 4 THWN #12 CU
- 2 RHW (covered) #8 CU
- 3 RHW (covered) #6 CU

Start by determining the area of each conductor from Chapter 9, Table 5.

Area of #12 THWN is 0.133 × 4 = .0532

Area of # 8 RHW covered = .0835 × 2 = .167

Area of #6 RHW covered = .1041 × 3 = .3123

Total area = .5325 square inches

Next, look in Chapter 9, Table 4 for Rigid Metal Conduit and find the first number larger than **.5325**

This is listed as 0.610 in the column for "Over 2 Wires - 40%"

This provides your answer, which 1 1/4 inch conduit.

Hopefully this chapter has helped you understand how you may be applying the NEC® code requirements to projects in the field. By visualizing the conditions described, the steps involved, and the codes required in various exam questions, you should be able to increase your percentage of correct answers.

Review and Appying Principles

QUIZ I—BASIC ELECTRICAL PRINCIPLES

1. Frequency is measured in which of the following:
   - **a.** Voltage
   - **b.** Watts
   - **c.** Hertz
   - **d.** None of the above

2. For electroplating power, which of the following generators should be used:
   - **a.** Separately excited
   - **b.** Delta system
   - **c.** Split phase
   - **d.** None of the above

3. Power factor is equal to which of the following:
   - **a.** Apparent power multiplied by true power
   - **b.** True power divided by apparent power
   - **c.** Watts multiplied by amps
   - **d.** None of the above

4. The combined opposition to current by resistance and reactance is which of the following:
   - **a.** "Z"
   - **b.** Impedance
   - **c.** Both of the above
   - **d.** None of the above

5. In a generator, voltage is produced by which of the following conditions:
   - **a.** Electrical pressure
   - **b.** Vibration
   - **c.** Magnetism
   - **d.** Cutting lines of force

6. A capacitor performs which of the following functions:
   - **a.** It opposes changes in voltage
   - **b.** It generators voltage
   - **c.** It creates changes in amperage
   - **d.** None of the above

7. Where the current flow through a conductor is greatest will be the point where the voltage drop is the lowest.
   - **a.** True
   - **b.** False

8. Of all of the materials used for insulation, which of the following has the highest electrical breakdown strength and the longest life:

    a. Rubber
    b. Cloth
    c. Impregnated paper
    d. Colored vinyl

9. A ground rod shall not be less than 8 feet long.

    a. True
    b. False

10. A three horse power motor is equivalent to which of the following in watts:

    a. 746 watts
    b. 2238 watts
    c. 3000 watts
    d. None of the above

11. Illumination is measured in which of the following units:

    a. Lumens
    b. Foot candles
    c. Light per square inch
    d. None of the above

12. Some copper conductors are tinned to help prevent chemical reaction:

    a. True
    b. False

13. Which of the following should be tested to check for voltage to ground:

    a. From the breaker to the grounding neutral
    b. From hot to neutral
    c. From a breaker to the cabinet
    d. All of the above

14. During one complete rotation of 360 degrees, a three-phase 6-pole AC 34 kVA alternator on a Y-connected system will have how many of the following rotations:

    a. 4
    b. 3
    c. 18
    d. 12

15. Which of the following terms is used to describe the inductive action, which causes current to flow on the outer surface of a conductor:

    a. Halo effect
    b. Inductance
    c. Skin effect
    d. Insulated par value

16. The wattage rating of a resistor determines its ability to absorb heat.

    a. True
    b. False

Review and Applying Principles

17. A wiring system has two lengths of wires, where wire A has a resistance of 10 ohms and wire B is three times the length of wire A with two times the cross-sectional area, which means that resistance of wire B is equal to which of the following:

   a. 30 ohms **b.** 20 ohms
   c. 15 ohms **d.** 10 ohms

18. The circuit conductors between service equipment and the final branch circuit overcurrent device are considered to be which of the following:

   a. Service laterals **b.** Feeder conductors
   c. Grounded conductors **d.** A wire assembly

19. The sum of series voltage drops equals the applied voltage.

   a. True **b.** False

20. If a transformer ratio is 20/1, then which of the following is true:

   a. For every 20 turns in the primary, there is one turn in the secondary.
   b. The primary voltage is 20 times greater than that of the secondary.
   c. For every 20 rotations in the secondary, there is one rotation in the primary.
   d. None of the above

21. When a switch on a 110 amp system is closed, the total resistance of the switch is equal to which of the following:

   a. 55 amps **b.** 220 volts
   c. 1100 ohms **d.** Zero

22. If a series circuit has unequal resistances throughout the circuit, then which of the following is true:

   a. The lowest resistance has the highest current.
   b. The lowest resistance has the highest voltage.
   c. The highest resistance has the highest voltage.
   d. The highest resistance has the highest current.

23. When reduced to the most nominal component, the smallest element of matter is which of the following:

   a. One ohm of electrical current **b.** An atom
   c. Electrons **d.** None of the above

24. The armature reaction in an alternator is magnetizing at lagging loads.
    a. True
    b. False

25. A wire secured at one end with the other end fastened to a pole under tension is known as which of the following:
    a. Messenger wire
    b. A service drop
    c. A guy wire
    d. None of the above

26. Which of the following is characteristic of inductance?
    a. It creates opposition to voltage.
    b. It opposes changes in current.
    c. It creates changes in current.
    d. None of the above

27. Magnetic lines of force are also known as which of the following:
    a. Flux
    b. Alternating currents
    c. Reactance
    d. None of the above

28. Electric motors operate on the principal of induction and repulsion.
    a. True
    b. False

29. In applications where high starting torque is required, DC series motors are generally used.
    a. True
    b. False

30. If a motor enclosure is designed to contain sparks or flashes that could ignite surrounding vapors or gases, then the enclosure is considered to be which of the following:
    a. Ventilated
    b. Explosion proof
    c. Explosion resistant
    d. Fire proof

31. The output of a 3-phase transformer is measured in which of the following units:
    a. Volts
    b. Amperes
    c. Volt-amps
    d. Watts

32. In order to change the rotation of a DC series motor, which of the following must be done:
    a. The motor must be converted to AC.
    b. F1 and F2 must be reversed.
    c. The capacitor leads must be reversed.
    d. The frequency must be altered.

Review and Appying Principles

33. Bonding does which of the following:
   a. Provides electrical continuity
   b. Regulates the voltage
   c. Controls the load on the system installation
   d. All of the above

34. A meter used to measure both AC and DC voltage and current is known as which of the following:
   a. An ohms meter
   b. A multimeter
   c. A resistance meter
   d. None of the above

35. Which of the following materials is used on conduits inside and outside of a box:
   a. Bushings
   b. Couplings
   c. Lock nuts
   d. All of the above

36. In a coil, the higher the level of self-inductance:
   a. The lower the level of resistance will be
   b. The longer the delay will be in establishing current through it
   c. The greater the level of flux produced will be
   d. None of the above

37. A negative charge will do which of the following:
   I. Interact with negative charges
   II. Interact with positive charges
   III. Cause a short in a positive system
   a. I & III
   b. I & II
   c. I only
   d. All of the above

38. If a circuit breaker is installed vertically and the handle is flipped up, then the circuit breaker will be which of the following:
   a. Off
   b. Unable to operate properly
   c. On
   d. Locked in place

39. A dielectric electrical element is which of the following:
   a. A capacitor
   b. A grounding conductor
   c. An insulator
   d. None of the above

40. The electrical property of conductors that allows them to oppose the free flow of current traveling over them is measured in which of the following:

    a. Amperes   **b.** Ohms

    c. Volts   **d.** Watts

41. Two equally valued resistors in series have a total resistance of which of the following:

    a. Twice as much as one of the resistors

    b. The sum of one

    c. The difference between each resistance

    d. None of the above

42. An eddy current can cause which of the following:

    a. Power fluctuation   **b.** Unusable frequency ranges

    c. An increase in core loss   **d.** All of the above

43. In a transformer, the side with more turns is which of the following:

    a. The high voltage side   **b.** The primary

    c. The secondary   **d.** The low voltage side

44. In order to adjust the voltage of a constant speed DC generator, the field voltage must be changed.

    a. True   **b.** False

45. The most power lost in the form of heat would come from which of the following:

    a. A short circuit   **b.** Resistance

    c. "Z"   **d.** None of the above

46. Two transmission wires create corona when which of the following exists:

    a. The wires have a high potential difference

    b. The wires are installed overlapping or too close together

    c. The wires are spaced too far apart

    d. None of the above

47. Inductance and capacitance are not considerations in a DC current for which of the following reasons:

    a. DC supply has no frequency

    b. DC supply carries power equall

    c. Both of the above

    d. None of the above

48. Rearranging the control components in an electrical diagram in order to simplify tracing of a circuit is an example of a schematic diagram.

    a. True   **b.** False

49. If a transformer has a voltage step-down from 117 volts to 6 volts then the ratio of the transformer is equal to which of the following:

    a. 19, 5 : 1   **b.** 1 : 19, 5
    c. 180 : 1   **d.** 3 : 1

50. If the power factor of a circuit is increased, then which of the following occurs:

    a. The active power is increased
    b. The reactive power is decreased
    c. Both of the above
    d. None of the above

51. In a 3Ø circuit, how many electrical degrees separate each phase:

    a. 3   **b.** 120
    c. 180   **d.** 360

52. Ground fault protection for personnel on a 120 volt single-phase circuit is based on which of the following:

    a. Ungrounded current between conductors
    b. Unbalanced current between the grounded and ungrounded conductor
    c. Overload in the current carrying conductor
    d. Excessive voltage in the grounding conductor

53. Electrical appliances shall be installed in parallel in order to make the operation of each appliance independent of the others.

    a. True   **b.** False

54. In order to calculate watt hours, which of the following formulas must be used:

    a. $E \times I \times T^2$ **b.** Amps ÷ Volts
    c. $E \times I \times T$ **d.** $24 \div (E \times I \times T)$

55. The peak value of AC voltage determines the breakdown voltage of insulation.

    a. True   **b.** False

56. The efficiency of electrical equipment is determined by which of the following:
   a. The amount of friction present
   b. The motor speed divided by the input power
   c. The power output divided by the input
   d. None of the above

57. Which of the following is not applicable to an AC system:
   a. It can be transformed.
   b. It develops eddy currents.
   c. It can interfere with communications lines.
   d. It is suitable for charging batteries.

58. In order to correct a low power factor, which of the following should be checked first:
   a. Resistance
   b. Watts
   c. Inductive load
   d. All of the above

59. Which of the following is used to control speed in a DC motor:
   a. Field winding
   b. A primary transformer
   c. Ground fault breaker
   d. A bonding jumper

60. In order to connect an ammeter in series, the circuit must be opened at the voltage source and connected to the meter at the other end.
   a. True
   b. False

61. In an alternator, frequency is determined by which of the following:
   a. Motor size
   b. Rotation speed of the armature
   c. Voltage
   d. All of the above

62. In a transformer, the voltage per turn in the primary relates to the voltage per turn in the secondary in which of the following ways:
   a. The primary is more
   b. The secondary is more
   c. They are the same
   d. None of the above

Review and Appying Principles 289

63. In a lampholder, which of the following is true of the grounded conductor:

   a. It would connect to the screw shell

   b. It is the lead wire

   c. It would be fire-retardant

   d. All f the above

64. An electrical lighting timer switch is generally connected in parallel with the lighting circuit it controls.

   a. True   **b.** False

65. Bonding safely conducts any fault currents which may arise.

   a. True   **b.** False

66. If a two-wire 120 volt circuit has one grounded conductor and supplies a single-phase 5 horsepower motor, then the required number of overloads is which of the following:

   a. Two, one in each conductor

   b. One in the grounded conductor

   c. One in the ungrounded conductor

   d. Zero, no overload is required

67. Conductors are materials that can be charged by rubbing them against other wires.

   a. True   **b.** False

68. An incandescent lamp filament is made of nickel.

   a. True   **b.** False

69. Which of the following conductor types is both moisture and heat resistant:

   a. THHN   **b.** THW

   c. XHHW   **d.** None of the above

70. Electromagnetic induction involves a substance becoming a magnet when it is placed next to another magnet.

   a. True   **b.** False

71. A metal halide lamp is also referred to as a quartz lamp.

   a. True   **b.** False

72. Lampholders installed in damp or wet locations shall be which of the following:
- **a.** Waterproof
- **b.** Insulated
- **c.** Brass
- **d.** None of the above

73. In an open circuit the resistance is equal to which of the following:
- **a.** 0.012 ohms
- **b.** 1 coulomb
- **c.** Zero
- **d.** Infinity

74. A resistor's ability to absorb heat is gauged in which of the following units:
- **a.** Watts
- **b.** Ohms
- **c.** Volts
- **d.** Amps

75. The power factor for a single-phase 50 amp 100 volt electric motor that draws 25 watts is which of the following:
- **a.** 20
- **b.** 0.20
- **c.** 0.025
- **d.** 0.005

76. Based on Ohm's law, the voltage of a system is equal to which of the following:
- **a.** Current multiplied by the wattage
- **b.** Current multiplied by the resistance
- **c.** Amperes divided by the resistance
- **d.** Amperes multiplied by the current

77. The rate of flow of the electrical charges through a conductor is which of the following:
- **a.** Watts
- **b.** Ohms
- **c.** Voltage
- **d.** Current

78. According the Kirchoff's current law, electrical current will do which of the following in parallel circuits:
- **a.** Combine in the parallel path
- **b.** Divide and flow through each parallel path
- **c.** Increase in voltage
- **d.** Experience a voltage drop

79. Watts are equal to which of the following:
- **a.** Voltage divided by amps
- **b.** Amps multiplied by voltage times the power factor
- **c.** Ohms multiplied by amps and divided by the power factor
- **d.** None of the above

80. The amperage of a three-phase system can be calculated as which of the following:
- **a.** Watts divided by the value of the voltage times 1.73
- **b.** 1.73 × watts divided by volts
- **c.** Volts times 1.73
- **d.** Power factor divided by the value of the voltage times 1.73

81. A centimeter is equal to which of the following:
- **a.** 2 1/4 inches
- **b.** Inches × 2.54
- **c.** 1.98 inches
- **d.** None of the above

82. Efficiency is calculated as output divided by input.
- **a.** True
- **b.** False

83. Circuits that are run in series:
- **a.** Have a total voltage of 8.92 per circuit
- **b.** Have a total resistance equal to the sum of all the resistors
- **c.** Must be connected to a bonding jumper
- **d.** All of the above

84. Fahrenheit temperature is converted from Centigrade by using which of the following equations:
- **a.** Dividing the Centigrade temperature by 1.8
- **b.** (Temp C × 1.8) + 32
- **c.** Multiplying the Centigrade temperature by 32 and adding 1.798
- **d.** None of the above

85. The temperature of the area surrounding a conductor is known as the ambient temperature.
- **a.** True
- **b.** False

ANSWERS

QUIZ 1—BASIC ELECTRICAL PRINCIPLES

1.	C	23.	B	44.	B	65.	A
2.	A	24.	A	45.	B	66.	C
3.	B	25.	C	46.	A	67.	B
4.	B	26.	B	47.	A	68.	B
5.	D	27.	A	48.	A	69.	D
6.	A	28.	A	49.	A	70.	B
7.	B	29.	A	50.	B	71.	B
8.	C	30.	B	51.	B	72.	D
9.	A	31.	C	52.	B	73.	D
10.	B	32.	B	53.	A	74.	A
11.	B	33.	A	54.	C	75.	D
12.	A	34.	B	55.	A	76.	B
13.	D	35.	C	56.	C	77.	D
14.	B	36.	B	57.	D	78.	B
15.	C	37.	B	58.	C	79.	B
16.	B	38.	C	59.	A	80.	A
17.	C	39.	C	60.	B	81.	B
18.	B	40.	B	61.	B	82.	A
19.	A	41.	A	62.	C	83.	B
20.	A	42.	C	63.	A	84.	B
21.	D	43.	A	64.	B	85.	A
22.	C						

Chapter 12
MASTER ELECTRICIAN SKILLS

Electrician's Exam Study Guide

There are a number of differences between the Journeyman test and the Master electrician exam. Generally speaking, the test for the Master electrician's exam has more questions, lasts longer and includes more advanced electrical fundamentals and formulas. We have already briefly touched on some of these theories, but now it's time to build on the basics.

A master electrician has to be capable of determining the proper size for grounding conductors. Grounding conductor sizes are provided in Table **[250-122]**. Branch circuit and feeder conductors are protected by fuses or a circuit breaker, and the size of the grounding conductor is based on the rating of the fuse or circuit breaker. Let's look at an example:

If a 150 ampere circuit breaker protects a set of #1/0 AWG copper feeder conductors with insulation rated at 758 degrees C, then the equipment grounding conductor shall not be smaller than which of the following:

a. #6 AWG copper

b. #8 AWG copper

c. #10 AWG copper

d. #12 AWG copper

In this problem, you start by looking up 150 amperes in Table **[250.122]**. Since there is no listing for 150 amperes, you have to use 200 amperes, and choose #6 copper.

When conductors that comply with Section **[310-4]** are run in parallel, and each set of conductors is run in a separate nonmetallic raceway, then an equipment grounding conductor is required with each circuit conductor in the raceway. Each equipment grounding conductors has to be sized based on the rating of the circuit or feeder overcurrent device as outlined in Section **[250.122(F)(1)]**. Look at the example below and determine the correct answer:

If a feeder is protected by a set of 300 amp fuses, and runs from one panel to another as two parallel sets of #1/0 AWG copper conductors with 758-degree C insulations, and terminates in separate nonmetallic raceways, and a copper equipment grounding conductor runs with each parallel set of conductors, then the minimum equipment grounding size shall be no less than which of the following:

a. #4 AWG

b. #6 AWG

c. #8 AWG

d. #12 AWG

Look up the 300 amp overcurrent device in Table **[250.122]** to find that it would be necessary to run #4 AWG copper grounding conductors in each of the raceways.

One of the next variables you are likely to find not only on the exam, but in the field as well, is **Voltage drop**. Voltage drop is the reduction in voltage between the source of power in an electrical circuit and a device that utilizes the power. Voltage drop is present in all electrical circuits powering any device and must be taken into consideration in circuit design. The NEC® sets guidelines for the maximum voltage drop that is allowed in branch circuits, conductors and feeders. Sections **[210.19(A)(1) FPN #1]** and **[215.2(A) FPN #2]** require branch circuit conductors and feeders individually to be sized to prevent a voltage drop larger than 3 percent at the farthest outlet of power or 5 percent combined.

Master Electrician Skills

Ohm's law is used to calculate current, potential difference, and resistance. The potential difference is the voltage drop. Alternating current continually reverses direction in a circuit at 60 hertz (60 cycles per second). The voltage drop in an alternating current (AC) circuit is the product of the current and the impedance (Z) of the circuit. The equation is:

$$I\,(\text{Current}) = \frac{V\,(\text{Voltage})}{R\,(\text{Resistance})}$$

$V = I \times R$ Voltage = Current × Resistance (Ohms)

$I = V \div R$ Current = Voltage ÷ Resistance (Ohms)

$R = V \div I$ Resistance (Ohms) = Voltage ÷ Current

The formula for voltage drop is based on single-phase or 3-phase systems:

SINGLE PHASE VOLTAGE DROP: $VD = 2 \times I\,(\text{Current}) \times R\,(\text{Resistance})$

THREE PHASE VOLTAGE DROP: $VD = 1.73\,(\text{for 3-phase}) \times I \times R$

Resistance in a conductor is represented by "K" multiplied by the length of a conductor divided by the cross-sectional area of the conductor, which is represented in circular mils by CM.

$K = 12.9$ ohms for copper

$K = 21.2$ ohms for aluminum.

K = Direct-Current Constant. "K" represents the DC resistance for a 1,000-circular mils conductor that is 1,000 feet long, at an operating temperature of 75 degrees C.

I = Amperes: The load in amperes at 100%

L = Distance: The distance of the load from the power supply. When calculating conductor distance, use the length of the conductor—not the distance between the equipment connected by the conductor.

CM = Circular-Mils: The circular mils of the circuit conductor as listed in NEC® Chapter 9, Table 8.

> ▶ **Test** tip
> In the field you would arrive at length by adding distance along the raceway route to the amount of wire sticking out at each end. On the exam, however, you will simply be given a distance, or the question will include the distance with wording like this: "a conductor that is 25 feet long."

The formula for calculating resistance is also based on single-phase or 3-phase.

SINGLE PHASE RESISTANCE: $2 \times K\,(\text{Resistance}) \times L\,(\text{length})$ 4CM

THREE PHASE RESISTANCE: $1.73 \times K\,(\text{Resistance} \times L\,(\text{length})$ 4CM

Current flows through the non-zero resistance of a practical conductor and produces a voltage across that conductor. In alternating current (AC) circuits, there is more opposition to flow of current because of an interaction between electric and magnetic fields. The opposition of current within the conductor is called **Impedance**. The DC resistance of a conductor depends on the conductor's length, cross-sectional area, type of material, and temperature. The impedance in an AC circuit depends on the spacing and dimensions of the conductors and the frequency of the current. Electrical impedance, like resistance, is expressed in ohms and opposes the current flow in a circuit. Electrical impedance is the vector sum of electrical resistance, capacity reactance, and inductive reactance as illustrated below:

$$Z = \sqrt{R^2 + X_L^2}$$

Z = Impedance (ohms)

R = Resistance (ohms)

X_L = Inductance Reactance (ohms)

Reactance is the part of total resistance that appears in AC circuits only. Like other resistance, it is measured in ohms. Reactance produces a phase shift between the electric current and voltage in the circuit. Reactance is represented by the letter "X." The two types of reactance are **Inductive reactance** and **Capacitive Reactance**.

If $X > 0$, the reactance is said to be inductive.

If $X = 0$, then the circuit is purely resistive; in other words it has no reactance.

If $X < 0$, it is said to be capacitive.

The relationship between impedance, resistance, and reactance is illustrated by the equation:

$$Z = R + jX$$

Z = impedance in ohms

R = resistance in ohms

X = reactance in ohms

"j" = an imaginary unit of $\sqrt{-1}$

Inductive reactance is the resistance to current flow in an AC circuit, due to the effects of inductors in the circuit. **Inductors** are coils of wire, typically wires that are wound on an iron core. Transformers, motors and fluorescent ballasts are the most common types of inductors. Inductance opposes a change in current in a circuit and creates a lag in the voltage in the circuit. When the voltage begins to rise in the circuit, the current does not begin to rise immediately; instead it lags behind the voltage. It is like when you turn the heat on in your car; the fan starts blowing right away but the warm air takes a minute to start coming out. The amount of lag depends on the amount of inductance in the circuit. In electrical formulas, inductive reactance is signified by X_L; capacitive reactance by X_C. The amount of impedance in a basic inductive or capacitive situation boils down to just the reactance. The reactance formula looks like this:

$$X = X_L - X_C$$

Inductive reactance (symbol X_L) is caused by the fact that an electrical current is accompanied by a magnetic field. A current that waivers is accompanied by a varying magnetic field, which creates an electromotive force that resists the changes in current. The more the current changes, the more an inductor resists it. The reactance present is proportional to the frequency, which is why there is zero reactance in DC. There is also a phase difference between the current and the actual voltage that gets applied. The formula for Inductive Reactance is:

$$X_L = \omega L = 2\pi L$$

$X_L =$ the inductive reactance measured in ohms

$\omega =$ the angular frequency, also known as angular speed, which is the measurement of how fast an object (in this case electricity) is rotating, measured in radians per second

$f =$ the frequency, which is measured in hertz

$L =$ the inductance, which is measured in henries. If the rate of change of current in a circuit is one ampere per second and the resulting electromotive force is one volt, then the inductance of the circuit is one henry.

$\pi =$ the symbol for Pi. Pi is a mathematical constant equal to approximately 3.1415926535897932.

Capacitive reactance (symbol X_C) exists because electrons cannot pass through a capacitor, but effectively alternating current (AC) can pass. There is also a phase difference between the alternating current flowing through a capacitor and the potential difference across the capacitor's electrodes. The potential difference is the voltage present between two points, or the voltage drop over an impedance, and represents the energy that would be required to move a unit of electrical charge from one point to the other against any electrostatic field that might exist. The formula for Capacitive Reactance is the following:

$$X_C = \frac{1}{\omega C} = \frac{1}{2\pi f C}$$

$X_C =$ the capacitive reactance, which is measured in ohms

$\omega =$ the angular frequency, which is measured in radians per second

$f =$ the frequency, which is measured in hertz

$C =$ the capacitance, which is measured in farads. A capacitor has a value of one farad when one coulomb of stored charge creates a potential difference of one volt across the capacitor terminals.

At this point, you might feel like you just completed an electrical engineering course. While the theories we just covered are not something you are going to have to use everyday in the field, they are the building blocks of how electrical energy moves, so you need to be able to understand them. A practice quiz is provided for you below so that you can exercise your math skills.

MASTER LEVEL QUIZ

1. If the source of supply is 125 feet away, the voltage drop for a 120 volt, 20 amp single phase circuit that supplies a 12 amp load and has circuit conductors with a combined resistance of 0.35 ohms would be which of the following:

   a. 42 volts
   b. 21 volts
   c. 4.2 volts
   d. 2.1 volts

   > ▶ **Test** tip
   > Start by writing out an Ohms formula, $E = I \times R$, and insert the appropriate values from the question.

2. If a 120 volt, 20 amp single phase feeder circuit supplies a 12 amp load and has circuit conductors with a combined resistance of 0.35 ohms, and the source of supply is 125 feet away, then the approved conductor size, based on the resulting voltage drop for this run, would be which of the following:

   a. 4.2 volts
   b. 42 volts
   c. 20 amps
   d. 36 amps

   > ▶ **Test** tip
   > This is not asking for the same answer as question #1. Once you have calculated the voltage drop, check to see if it meets the requirements of Section [210.19] FPN #4. Hint–3% maximum

3. A 2 amp DC circuit which reads 10 volts on a voltmeter has resistance in the circuit equivalent to which of the following:

   a. 20 ohms
   b. 12 ohms
   c. 5 ohms
   d. 2 ohms

4. If a single-phase 60 amp series circuit that has resistance in the circuit of R1 at 10 ohms, R2 at 5 ohms and R3 at 15 ohms, then the total of the voltage drop across each resistance is equal to which of the following:

   a. 30 volts
   b. 15 volts
   c. 1800 amps
   d. 3600 amps

5. If a single-phase 60 amp series circuit that has resistance in the circuit of R1 at 10 ohms, R2 at 5 ohms and R3 at 15 ohms, then the voltage of the circuit is equal to which of the following:

   a. 15 volts
   b. 30 volts
   c. 60 volts
   d. 120 volts

6. In a single-phase resistance style 120 volt heater rated at 4500 watts, the amount of amperes used by the unit is which of the following:

   a. 4.5 amps
   b. 15.5 amps
   c. 37.6 amps
   d. 45 amps

7. In an apartment building with a three-phase, four-wire, 120/208 electrical service with a single-phase, three-wire 120/208 feeder where two ungrounded and one neutral conductor runs to each dwelling unit, only the 120 volts loads are operating and the loads are balanced. If the current to each ungrounded conductor is 42 amperes, then the current of the neutral conductor to one dwelling unit is equal to which of the following:

   a. 0 amps
   b. 21 amps
   c. 42 amps
   d. 84 amps

8. In an electrical conductor that consists of 16 strands, each with a diameter 0.0837 inches, the area of the conductor would be which of the following:

   a. 83.7 circular mils
   b. 83700 circular mils
   c. 70057 circular mils
   d. 112091 circular mils

9. Assume that a three-phase, 460 volt continuous load wound induction motor is rated at 40 horsepower, and the nameplate indicates the full load current is 45.5 amps and a secondary full load current of 82 amps. The motor is rated at a temperature rise of 40 degrees C. Based on the fact that the resistor bank is separate from the controller and is classified at "medium intermittent duty," then the minimum current rating of the secondary conductors running between the motor and the controller must be no less than which of the following:

   a. 41 amps
   b. 46 amps
   c. 52 amps
   d. 65 amps

10. Building "A" is supplied with 120/240 volt single-phase power from building "B" on the same property. "B" is supplied from a three-wire with two ungrounded conductors and one neutral without any metallic water pipe or other metal equipment connections to the building or equipment ground fault protection installed. The neutral conductor must:

    a. Not be connected to a grounding electrode in the second building
    b. Not be connected to a grounding electrode in either building
    c. Be bonded to the disconnect enclosure in the second building and connected to a grounding electrode
    d. Tie into the ungrounded conductor of one of the buildings with a grounding jumper

11. A feeder runs from one part of a building to another under the floor in two parallel sets of rigid non-metallic conduits with type RHW copper conductors size AWG #500 and is protected by 800 ampere fuses. It must have a minimum size copper equipment grounding conductor in each conduit run of no less than which of the following:

- **a.** 1/0 AWG
- **b.** 20 AWG
- **c.** 10 amps
- **d.** 5 AWG

12. A surge arrestor for a 480 volt electrical system requires a connecting conductor that is #14 copper or larger.

- **a.** True
- **b.** False

13. A copper THHN feeder conductor consists of 40 amperes of continuous load and 35 amperes of non-continuous load, and supplies a load that does not contain any general use receptacles but has three current-carrying conductors in a raceway with terminations rated at 75(degrees C. This means that the minimum standard overcurrent device required for the feeder is which of the following:

- **a.** 70 amps
- **b.** 76 amps
- **c.** 86 amps
- **d.** 90 amps

14. If a feeder conductor contains type RHW 500 kcmil copper conductors protected by a 400 amp circuit breaker, and a tap is made to the feeder in order to supply a 100 amp main circuit breaker 22 feet from the tap location, then the minimum size copper type THWN conductor that is permitted between the tap and the panelboard is which of the following:

- **a.** 1/0 AWG
- **b.** 2 AWG
- **c.** 3 AWG
- **d.** 4 AWG

15. If a single family dwelling unit has 2680 square feet of living space and all of the 120 volt general illumination circuits are rated at 15 amperes, then the minimum number of circuits required is which of the following:

- **a.** 5
- **b.** 6
- **c.** 7
- **d.** 8

16. A single family dwelling with 2400 square feet of living area has a 120/240 volt three-wire electrical service and contains the following: a 3.5 kW 240 volt electric water heater, a 12 kW electric range, a 1.5 horsepower 240 volt central air conditioner, a 1/2 horsepower 120 volt garbage disposal, a 1/3 horsepower 120 volt furnace blower motor, a 1.2 kW 120 volt dishwasher, and a 5 kW clothes dryer. The total minimum load required for the general illumination, small appliances, and laundry without any demand factors would be which of the following:

- **a.** 1500 VA
- **b.** 7200 VA
- **c.** 11700 VA
- **d.** 15000 VA

17. If a three-phase 75 kVA transformer is connected to the primary at 480 volts and 120/208 volts on the secondary, then the full load current of the transformer secondary would be which of the following:

- **a.** 240 amps
- **b.** 208 amps
- **c.** 90 VA
- **d.** 25 VA

18. In order to correct the power factor, a three-phase, 480 volt, 92 kVAR capacitor bank located 6 feet from the main service of a 3200 square foot office building has a minimum required ampere rating for the conductors in the capacitor bank of which of the following:

▶ **Test** tip
Underline only the pertinent elements of the question to use in your calculations.

- **a.** 240 amps
- **b.** 180 amps
- **c.** 150 amps
- **d.** 110 amps

19. A recreational vehicle located on one of 20 RV sites is connected to a 20 amp 125 volt receptacle, which must also be which of the following:

- **a.** 6.5 feet off of the ground
- **b.** Meet a 47% demand factor
- **c.** Connected to a bonding jumper
- **d.** None of the above

20. In an eight unit apartment building, each kitchen contains a 3.5 kW 240 volt electric range. The demand load for the electric service to the building must include an allowance for the minimum demand load of all of the ranges, which would be which of the following:

- **a.** 28 kVA
- **b.** 21 kVA
- **c.** 14 kVA
- **d.** 7.5 kVA

21. A retail store has 3000 square feet and 30 feet of show window. The service is a 120/240 volt single phase 3-wire service, and there is an actual connected lighting load of 8500 VA. There are a total of 80 duplex receptacles. Given these facts, the total calculated load is which of the following:

- **a.** 9000 VA
- **b.** 12200 VA
- **c.** 16200 VA
- **d.** 28400 VA

22. The same retail store with 3000 square feet and 30 feet of show window, a 120/240 volt single phase 3-wire service, an actual connected lighting load of 8500 VA, and a total of 80 duplex receptacles would require a minimum standard feeder overcurrent protection size of which of the following:

    a. 480 volt   **b.** 240 volt

    c. 135 amp   **d.** 150 amp

23. A multi-family dwelling has 40 units, with 208Y/120 volt three-phase main service and two-phase legs and neutral running to two meter banks of 20 units each. One half of the units are equipped with 12 kW electric ranges rated in accordance with the requirements of NEC® Section [220.55], and the other units have gas ranges. The only laundry connections are in a laundry room in the basement, available to all of the tenants. Each apartment unit has 840 square feet. Given these conditions, the net calculated feeder neutral load is which of the following:

    a. 64700 VA   **b.** 34500 VA

    c. 276.5 amps   **d.** 253.6 amps

24. A multi-family dwelling has 20 units, with 120/240 volt single phase main service running to two meter banks of 20 units each. One half of the units are equipped with 12 kW electric ranges rated in accordance with the requirements of NEC® Section [220.55], and the other units have gas ranges. The only laundry connections are in a laundry room in the basement, available to all of the tenants. Each apartment unit has 840 square feet. Given these conditions, the minimum feeder size for units without electric ranges would be which of the following:

    a. 49.5 amps   **b.** 16.2 amps

    c. 12 amps   **d.** 5 amps

25. A project requires the installation of twelve 1.4 ampere, 120 volt, fluorescent lights fixtures on two 20 amp branch circuits, as well as three 120 volt, 5.6 ampere electric fans on individual circuits in a building with a 120/240 single phase three-wire electric service. The minimum neutral current allowed for these loads is which of the following:

    a. 0 amps   **b.** 5.6 amps

    c. 16.8 amps   **d.** 33.6 amps

ANSWERS

MASTER LEVEL QUIZ

1. **C** $I = 12$ amps, $R = 0.35$ ohms
2. **D** The voltage drop of the 20 amp conductors listed in the question is 3 1/2 percent. Section [210.19] FNP #4 requires a maximum of 3%. 3 percent of 120 volts is a 36 amp conductor size.
3. **C** $R = E \div I$; $E = 10$ and $I = 2$
4. **B** $E_1 = 0.5 \times 10$ volts; $E_2 = 0.5 \times 5$ volts; $E_3 = 7.5$ volts
5. **C** $E = 2 \times I + R$; $2(10 + 5 + 15) = 60$
6. **C** 4500 watts ÷ 120 volts = 37.6 amps
7. **C** Reference Section [310.15(B)(4)(b)]
8. **D** First you need to convert the 0.0837 inches to mils, which is 83.7. Because you are dealing with circular mills, you have to multiply that times 2, which gives you 7000.57 circular mills. Next, multiply that by the number of strands in the conductor $16 \times 7005.7 = 112091.2$ (rounded down to 112091).
9. **D** 52 amps × 1.25 = 65 amps. Refer to the full load current listed in Table [430.150], not the current listed on the nameplate; this requirement is found in Section [430.6(A)(1)]. Then multiply that value by 1.25 as required by Section [430.22(A)].
10. **C** Reference Section [250.32(B)(1)]
11. **A** Reference Table [250.122]
12. **A** Reference Section [280.21]
13. **D** Reference Sections [215.3] and [240.6]; (40 amps × 1.25) + (36 amps × 100) = 86 amps. Since 86 is not an industry standard size, you need to choose 90 amps.
14. **A** Reference: see the 25 foot tap rule in Section [240.21(B)(2)]. Since the tap is ahead of the 100 amp overcurrent device and is only protected by the 400 amp CB, the tap cannot have an ampacity less than 1/3 that of the feeder overcurrent device. 400 amps ÷ 3 = 133 amps. Next, look up the feeder amperage and permitted wire size for type RHW listed in Table [310.15(B)(6)].
15. **A** Reference Sections [210.11(A)] and [220.423(A)] for minimum load requirements. Divide the load by 120 volts, then divide this value into the circuits to determine the number of circuits required. 3 VA ÷ 2680 feet 2 = 8040 VA; 120 volts × 15 amps (per circuit) = 1800; 8040 ÷ 1800 = 4.5 (rounded up to 5).
16. **C** Reference Table [220.3(A)] to see that 3 VA is required for every square feet of living area. 2400 square feet × 3 VA = 7200 VA. The small appliance load in Section [220.11(C)(2)] is 1500 VA and comes to 2 small appliance loads. 1500 VA × 3 = 3000; Section [210.11(C)(2)] requires 1500 VA for a dwelling laundry circuit. 7200 + 3000 + 1500 = 11700 VA.
17. **B** Use the formula $C_{secondary} = \dfrac{75\text{kVA} \times 1000}{1.73 \times 208\text{v}} = 208$ amps

18. **C** The first step is to calculate how much current will be going to the capacitor bank once it is energized, which is the same formula you would use to calculate the full load current of a transformer, except you need to use kVAR's instead of kVA.

$$C\ capacitor = \frac{92\,kVAR \times 1000}{1.73\,(3\,phase) \times 480\,(volts)} = 110.8\ amps$$

Next you need to find the minimum ampere rating of the conductors in the capacitor bank using the requirement in Section [460.8(A)], and you will find that the ampacity cannot be more than 135% (which is 1.35).

Multiply the capacitor current of 110.8 × 1.35 = 149.58 rounded up to 150 amps.

19. **A** Reference Section [551.77(D)]. Note: we thought your mind could use a quick break at this point.
20. **B** Reference [Section 220.17] lists a demand factor of 0.75 for four or more appliances.
    8 ranges × 3.5 kW each = 28 × 0.75 demand factor = 21 kVA
21. **D** Reference Chapter 9, Annex D, Example D3
22. **D** Reference Chapter 9, Annex D, Example D3. NOTE: the question is asking for the minimum standard feeder overcurrent protection size.
23. **D** Reference Chapter 9, Annex D, Example D5(a)
24. **B** Reference Chapter 9, Annex D, Example D4(a)
25. **A** Balance the load by placing 4 fixtures and 2 fans on one circuit and 8 fixtures and one fan on the other as shown below:

LEG 1		LEG 2	
4 Fixtures	5.8 amp	8 Fixtures	11.2 amps
Fan	5.6 amps	Fan	5.8 amps
Fan	5.6 amps		
	16.8 amps		16.8 amps

Chapter 13
TIMED TEST 1

Electrician's Exam Study Guide

Now you are ready to take a test. This is a timed, open book test; your goal is to answer all fifty questions in two hours. Write down the time you start taking the test and the time you finish. If you have to stop along the way, then write down your pause time and your restart time. You want to make sure to practice taking this test and answering all of the questions within the two hours allotted. After you have completed the test, checked your answers. The answer section provides many test tips and references to sections in the NEC® so that you can go back through the sections and review any material you answered incorrectly.

> ▶ **Test** tip
>
> Since this is a timed test, skip over any questions that you just don't know the answers to and come back to them. The material in other test questions may help to point you to the answers you originally didn't know. No matter what, answer all the questions before you complete the test, even if it means using your "best guess." Remember, incorrect answers do not count against your score, and who knows-you may just guess correctly.

TIMED TEST I

TIME ALLOWED: 2 HOURS

1. Devices are equipment that carry current, but do not perform which of the following functions?
    - a. Serve a grounding function
    - b. Utilize electric energy
    - c. Provide overcurrent protection
    - d. Control power

2. Which of the following forms of electrical equipment is not considered a device?
    - a. Receptacle
    - b. Three-way switch
    - c. Light bulb
    - d. Disconnect switch

3. In construction of one- and two-family dwellings, the electrical circuits are designed by which of the following professionals?
    - a. Electrician
    - b. Architect
    - c. Design Engineer
    - d. Electrical Inspector

4. In the National Electrical Code, which of the following words means that "it must be done?"
    - a. May
    - b. Shall
    - c. Should
    - d. Recommended

5. Which of the following is not included when calculating the usable area of a dwelling and computing the required lighting load?
   a. Living room
   b. Crawl space
   c. Finished basement
   d. Bathroom

6. Wire sizes larger than 4/0 AWG are expressed in which of the following?
   a. Metric units
   b. kcmil
   c. Inches
   d. Radians

7. In general, residential loads and devices such as lighting fixtures and receptacles are connected to branch circuits defined by the *National Electrical Code®* in series.
   a. True
   b. False

8. Which of the following colors of insulation or marking are permitted to identify "hot" phase conductors in conduit wiring methods?
   a. Black, red, and blue for 120/208-volt systems
   b. Yellow, orange, and brown for 277/480-volt systems
   c. Any colors except white, gray, or green
   d. All of the above

9. Wiring devices for use on 120-volt nominal circuits are marked as which of the following?
   a. 110 volts
   b. 120 volts
   c. 125 volts
   d. All of the above

10. If an existing two-wire, nongrounding receptacle is replaced in a location where the *Code* requires a GFCI receptacle, it must be replaced with which of the following?
    a. GFCI receptacle
    b. Two-wire, nongrounding-type receptacle
    c. Three-wire, grounding-type receptacle
    d. All of the above

11. Two resistors connected in a series across a 120 volt power source, where the first has a resistance of 28 ohms and the second has a resistance of 7 ohms, will generate a voltage drop across the first 28 ohms resistor of which of the following?
    a. 96 volts
    b. 60 volts
    c. 28 volts
    d. 120 volts

12. Based on the diagram below, a 4 ohms resistor, a 6 ohms resistor and a 12 ohms resistor are all connected in parallel. The total resistance in the circuit will be which of the following?

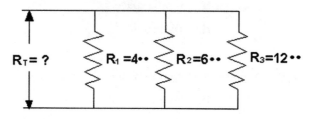

   a. 22 ohms
   b. 12 ohms
   c. 4 ohms
   d. 2 ohms

13. Electrical motors with the highest voltage rating are used for which of the following purpose(s)?
   a. To produce the maximum possible power from the motors
   b. To lower power consumption
   c. To reduce the size of the supply conductors required
   d. All of the above

14. Two incandescent 100 watt lamps operating for 8 hours at an average electrical energy cost of .12 cents per kilowatt hour will have a total energy cost equal to which of the following?
   a. $1.92
   b. $1.16
   c. $0.32
   d. $0.19

15. In a dry-type transformer with a primary to secondary turns ratio of 4:1, if the primary is rated at 480 volts, then the secondary voltage will be which of the following?
   a. 1200 volts
   b. 480 volts
   c. 240 volts
   d. 120 volts

16. In a single phase, 3-wire 120/240 volt service entrance, where the hot conductor current is 28 amperes on the first leg and 42 amperes on the second leg, the current in the neutral will be which of the following?
   a. 14 amperes
   b. 56 amperes
   c. 84 amperes
   d. 126 amperes

17. Based on the diagram options below, which figure illustrates a 3-phase electrical system where one ungrounded conductor has a higher voltage to the ground than the other hot conductor?

Figure A

Figure B

Figure C

Figure D

18. The total resistance of the circuit conductors for a single-phase load with a copper conductor resistance of 0.510 ohms/k ft. that is 125 feet from the current supply is which of the following?

   a. 1.255 ohms
   b. 0.5100 ohms
   c. 0.2550 ohms
   d. 0.1275 ohms

19. If the diameter of a solid copper conductor is 0.125 inches, then the area of the conductor is equal to which of the following?

   a. 15,625 cmil
   b. 12,265 cmil
   c. 31,250 cmil
   d. 12,500 cmil

20. Which of the following conductors has the largest cross-sectional area in size?

   a. 14 AWG
   b. 4 AWG
   c. 6 AWG
   d. 12 AWG

21. According to the *National Electric Code,* which of the following must be true of wiring installed to meet the minimum *Code* requirements?

   a. It must provide the most convenient wiring applications
   b. It is safe and free from hazards
   c. It must allow for future expansion needs
   d. All of the above

22. When a 120/240 volt 100 amp panelboard with a removable front cover is installed in a single-family dwelling, and the opposite wall in front of the panelboard is concrete block construction, the minimum allowable distance between the front of the panelboard and the concrete block wall is which of the following?

   a. 18 inches
   b. 24 inches
   c. 36 inches
   d. 42 inches

23. On a dwelling kitchen counter, the distance from any point along the wall line to a receptacle may not exceed which of the following?

   a. 12 inches
   b. 18 inches
   c. 24 inches
   d. 30 inches

24. A single-family dwelling unit sits at grade level and has one bathroom, an attached garage, and an unfinished basement with a furnace, washer and dryer, and the electrical service panel. Based on one ground-fault circuit-interrupter device used to protect all of the receptacles on the GFCI circuit, what is the minimum number of 125 volt GFCI devices required in this dwelling?

   a. 3
   b. 4
   c. 5
   d. 7

25. The lowest point of the drip loop of a 208/120 volt, 3-phase, 4-wire electrical system shall have a minimum clearance above a pedestrian sidewalk which is no less than which of the following?

- **a.** 10 feet
- **b.** 12 feet
- **c.** 12 1/2 feet
- **d.** 16 feet

26. If there are three 500 kcmil THHN conductors in a raceway at 70 degrees F and terminated at equipment rated at 90 degrees C, then the maximum allowable current for each of the conductors is which of the following:

- **a.** 447 amps
- **b.** 430 amps
- **c.** 500 amps
- **d.** 600 amps

27. A 24 inches long conduit nipple may be filled to a percentage of its total cross sectional area, and that percentage is which of the following:

- **a.** 60 percent
- **b.** 40 percent
- **c.** 25 percent
- **d.** 10 percent

28. Unopened drums of gasoline purchased for use at a bulk storage plant may be stored in which of the following outdoor areas:

- **a.** Class II, Division 1
- **b.** 25 feet away from a building or structure
- **c.** In unclassified areas
- **d.** None of the above

29. A light fixture may be installed in a clothes closet if there is a clearance of 18 inches maintained from the storage area.

- **a.** True
- **b.** False

30. Except where it is used in cables, the minimum thickness of sealing compound used in a conduit seal shall be no less than which of the following:

- **a.** 5/8 inch
- **b.** 1/2 inch
- **c.** 1/4 inch
- **d.** 1/8 inch

31. A three-phase squirrel cage motor operates at 400 volts and 65 amps when running at full load, which means that the horsepower rating for this motor based on the NEC® is which of the following:

- **a.** 65 horsepower
- **b.** 50 horsepower
- **c.** 40 horsepower
- **d.** 30 horsepower

32. Junction boxes do not have to be accessible if they are installed in a fire-rated wall or ceiling.

    a. True   **b.** False

33. Flexible metallic tubing may be used in runs not to exceed 10 feet in length.

    a. True   **b.** False

34. A fixture supported by the screw shell of a lampholder may not exceed which of the following:

    a. 2 1/2 pounds   **b.** 4 pounds

    c. 6 pounds   **d.** 10 pounds

35. The disconnection means for a 2300 volt motor must be capable of being locked in the open position.

    a. True   **b.** False

36. The frame of an electric range may be considered a grounding source if it is connected to the grounded conductor of a 120/240 volt branch circuit which is not less than which of the following:

    a. #6 copper   **b.** #10 copper

    c. #8 aluminum clad   **d.** None of the above

37. The motor branch circuit fuse size for a general motor application must be determined by referring to which of the following:

    a. NEMA standards   **b.** NEC® tables

    c. NEC® FPNs   **d.** The motor nameplate listing

38. The grounding electrode conductor connection to a driven ground rod shall be which of the following:

    a. Made with wire taps

    b. Accessible

    c. Direct-buried

    d. Not be required to be accessible

39. A motor conductor shall open all conductors to a motor at the same time.

    a. True   **b.** False

40. Edison-base fuses are acceptable for new single-family and two-family dwellings.

    a. True   **b.** False

41. Individual open service conductors for 600 volts or less that are installed in dry locations must be separated by which of the following:

   a. A tile or concrete barrier

   b. A distance of 2 1/2 inches of clear space

   c. A permanent divider that provides 4 inches of separation

   d. None of the above

42. Photovoltaic system currents are considered to be which of the following:

   a. Inverted currents   **b.** Continuous currents

   c. Limited currents   **d.** Stand alone systems

43. Lighting that is required for a hoist pit shall meet which of the following requirements:

   a. Connections must be run in flexible metal conduit

   b. It will not be connected to the load side of the ground-fault circuit interrupter

   c. The switch will not be readily accessible

   d. All of the above

44. Low voltage equipment in a hospital that comes in frequent contact with people's bodies must be which of the following:

   a. Able to operate at an electrical potential of 10 volts or less

   b. Approved as intrinsically safe

   c. Be moisture resistant

   d. All of the above

45. An AC general use snap switch must not be used to control a tungsten filament lamp load.

   a. True   **b.** False

46. If a 240 volt conductor is 500 feet long and supplies a 10 amp load and the conductor has a resistance of 1.45 ohms per 1000 feet, then the total voltage drop in the conductor circuit conductors would be which of the following:

   a. 10 percent   **b.** 7 percent

   c. 3 percent   **d.** 1.45 percent

47. Grounding equipment with a 40 amp automatic overcurrent device in the circuit ahead of the equipment requires a minimum size copper equipment grounding conductor of which of the following sizes:

   a. #8   **b.** #10

   c. #12   **d.** None of the above

48. An outlet box mounted in a combustible wall must meet the following requirement:
- **a.** Be set back 1/2 inch from the finished wall surface
- **b.** Be recessed 1/4 inch from the finished wall surface
- **c.** Be flush with the finished wall surface
- **d.** Be a metal box attached with 1/4 inch screws

49. Flexible cords shall only be permitted to be used to supply portable appliances.
- **a.** True
- **b.** False

ANSWERS

TIMED TEST 1

1. B
2. C
3. A

> ▶ **Test** tip
> Although different professionals can and do design electrical circuits, remember to always pick the BEST, most accurate answer.

4. B
5. B
6. B
7. B   Defined in parallel
8. D
9. C
10. A
11. A

> ▶ **Test** tip
> The voltage drop will be in the same proportion as each resistor in a series is to the total circuit resistance. 28 ohms is 80 percent of the total resistance (35 ohms, which is 28 + 7), so 80 percent of the voltage will be across the 28 ohms resistor. 0.8 × 120 volts = 96 volts

12. D $\dfrac{1}{R_T} = \dfrac{1}{4} + \dfrac{1}{6} + \dfrac{1}{12} = \dfrac{6}{12}$ $R_T = \dfrac{12}{6} = 2\ \text{ohms}$

13. C

> ▶ **Test** tip
> Check the current of a particular motor at two different voltages in Table [430.148] or Table [430.150].

14. D   100 watts × 2 bulbs = 200 watts or 0.2 kW
       0.2 kW × 8 hours = 1.6 kilowatt hours
       1.6 kWh × $0.12 = $0.19

15. D $\dfrac{\text{Primary Voltage}}{\text{Turns Ratio}} = \text{Secondary Voltage}$ $\dfrac{480\,\text{v}}{4} = 120\ \text{volts SDF}$

16. **A**

> ▶ **Test** tip
>
> In a single-phase, 3-wire system, the current on the common neutral is the difference in the current flowing on the ungrounded legs. 42A − 28A = 14 amperes.

17. **B**

> ▶ **Test** tip
>
> The question describes a 4-wire delta. The voltage from the neutral is connected to the center of one of the transformers at 208 volts to the phase at the top of the delta.

18. **D**

> ▶ **Test** tip
>
> There are two wires in the 125-foot run. 2 × 125 = 250, which is 25 percent of 1000 feet (k/ft.). 0.25 × 0.510 = 0.1275 ohms

19. **A**

> ▶ **Test** tip
>
> 0.125 inches = 125 mils   125 mils × 125 miles = 15,625 cmil

20. **B** Article 310, Table [310.16]
21. **B** Section [90.1(A)]
22. **C** Article 110, Table [110.26(A)(1)] column
23. **C** Section [210.52(C)(1)]
24. **B**

> ▶ **Test** tip
>
> A single-family dwelling assumes there is a kitchen, even though it is not spelled out in the description. Section [210.8(A)] states that 2 circuits are required for kitchen counters. One GFCI is needed for each circuit per Section [210.52(B)(2)]. One circuit is required for the bathroom receptacles, and one GFCI, as listed in Section [210.11(C)(1)]. The garage, basement, and outdoor receptacles can all be on one circuit with one GFCI. The washer, dryer and furnace do not require individual GFCI devices.

25. **A** Section [230.24(B)(1)]
26. **A** Table 310.6 plus temperature derating factors, and Section [110.14].

> ▶ **Test** tip
>
> The correction factor at the bottom of Table [310.6] for 70 degrees F is 1.04 for insulated conductors rated at 90 degrees C, such as THHN. Without the derating factor, the answer would have been 430. Utilizing the derating factor, you need to multiply the derating factor 1.04 x 430 amps = 447 amps.

27. **A** Chapter 9 Table 1 Note 4

> ▶ **Test** tip
>
> Where conduit nipples or tubing nipples have a maximum length not to exceed 24 inches and are installed between boxes, cabinets, or similar enclosures, the nipples shall be allowed to be filled to 60 percent of the total cross sectional area.

28. **C** Table 515.3 provides this permission for bulk storage plants
29. **A** Section [410.8(A)]
30. **A** Section [501.15(C)(2)]
31. **B** Table [430.250]
32. **B** Section [314.29] requires that wiring in junction boxes be accessible
33. **B** Section [360.12(6)] prohibits lengths over 6 feet
34. **C** Section [415.15(A)]
35. **A** Section [430.227]
36. **B** Section [250.140(2)]
37. **B** Sections [430.6(A)(1)] and [430.52(C)(1)]
38. **D** Section [250.68(A)] Exception
39. **B** Section [430.84]
40. **B** Section [240.51(B)] states that Edison-base plug fuses can only be used to replace existing fuses of that type
41. **B** Table [230.51(C)]
42. **B** Section [690.8(B)]
43. **B** Section [620.24(A)]
44. **D** Section [517.64(A)(1) & (2)]
45. **B** Section [404.14(A)]
46. **C** General Electrical Knowledge of $E = I \times R$ $E = 10 \times (500 \div 1000) \times 1.45$, which equals 7.25 volts. The percentage of the voltage drop is $7.25 \div 240 \text{ volts} \times 100 = 3 \text{ percent}$
47. **B** Table [250.122]
48. **C** Section [314.20]
49. **B** Section [400.7(A)]

Chapter 14
TIMED TEST 2

Electrician's Exam Study Guide

This test is also timed. Many of the licensing exams have between 50 and 100 questions, and you are allotted 3 hours or more for the test. Here you are permitted 2 hours for the practice test, which comprises 50 questions. The test is an open book exam. You are allowed to use your NEC® book and a calculator. This is an opportunity to set aside two full hours and take the test from start to finish. Remember the testing tips you have learned so far.

- Read the question carefully in order to understand the meaning of the question and determine which answer option is the most correct.

- If you need to, circle key words or values in the example so that you can break the question down into clearer components.

- Remember, incorrect answers don't count against you. It is in your best interest to answer every question.

- Some questions ask for the answer that is NOT a requirement or code.

- When reading an exam question that describes **Minimum** allowable installations, look for key words and phases such as: at least, no less than, and smallest allowable. For questions describing **Maximum** allowable requirements, look for key words or phases such as: no greater than, most, greatest, highest, and shall not exceed.

- An answer option of "Either of the above" is very different from "Both of the above". Both of the above ties the two answer options together and requires that **both** conditions must exist. Either of the above means that one or the other could be true.

- Eliminate extra facts or descriptions that have nothing to do with the answer. If a question requires you simply to add amperes, then the voltage of the system, square footage of the building or color of the light fixtures has nothing to do with picking the correct answer. Don't be distracted by superfluous information.

- Pay close attention to whether a system described in the test question is single phase or three-phase and remember to use the factor of 1.73 for three-phase calculations.

- If you are really struggling with a question, move on and come back to it later. Often in the course of answering other questions, the solution will become clear.

Now you are ready to begin. The answer sheet for this test provides explanations and code references to help you understand which answers are correct and why; it also enables you to go back and study specific areas in the code that you may have missed. Let's get started. Check the clock; your two hours start now.

TIMED TEST 2

TIME ALLOWED: 60 MINUTES

1. Size #14 AWG, type THHN cable is permitted in for use in wet locations.

   a. True   **b.** False

2. Branch circuit conductors of 600 volts nominal or less must have an ampacity not less than the minimum load that is to be served.

   a. True   **b.** False

3. Type AC cables are allowed for which of the following uses:

   a. In concealed work

   b. In wet locations

   c. For aerial service power

   d. None of the above

4. Enclosures around electrical equipment shall be considered accessible if they meet which of the following conditions:

   a. Are controlled by combination locks

   b. Are surrounded by fencing that is 7 feet tall

   c. Both of the above

   d. None of the above

5. If the length in a ground return path does not exceed 8 feet, then flexible metal conduit and tubing is permitted for equipment grounding.

   a. True   **b.** False

6. The connection between the grounded circuit conductor and the equipment grounding conductor at the service is which of the following:

   a. A main bonding jumper

   b. A grounding connector

   c. Service grounded conductor

   d. None of the above

7. Secondary overcurrent protection is not required for fire pump wiring installations.
   a. True
   b. False

8. A disconnecting means for a controller is also permitted to be the disconnecting means for the motor and driven machine, even where out of sight of the motor and driven machine:
   a. in an industrial building with qualified service personnel, if capable of being locked in the open position
   b. if capable of being locked in the open position
   c. in commercial and industrial buildings with qualified service personnel
   d. if provided with an alarm to indicate when the disconnect is closed

9. The computed load of a feeder circuit shall not be less than the sum of the branch circuit loads after any applicable demand factors have been applied.
   a. True
   b. False

10. The conductors connecting to motor controllers and to control devices are required to be:
    a. copper, aluminum, or copper-clad aluminum conductors
    b. only copper-clad aluminum conductors
    c. copper or copper-clad aluminum conductors
    d. only copper conductors unless identified for other materials

11. AC - DC general use snap switches may be used to control inductive loads that do not exceed the rating at the voltage of which of the following:
    a. 5 percent
    b. 45 percent
    c. 50 percent
    d. 100 percent

12. A 20 amp laundry is always required in a dwelling unit.
    a. True
    b. False

13. A hospital's essential electrical system is comprised of an emergency system and which of the following systems:
    a. A life safety system
    b. An addressable fire alarm system
    c. Both of the above
    d. None of the above

14. Shore power receptacles must be rated at a minimum of 30 amps.

   a. True
   b. False

15. For hospital grade receptacles installed in a patient bed location in a critical care unit, which of the following must be true:

   a. At least one of them must be connected to either the normal system branch circuit or an emergency system branch circuit
   b. The branch circuit must be supplied by a different transfer switch than the rest of the receptacles at that bed location
   c. All of the above
   d. None of the above

16. If a transformer has a primary winding of 4 turns and a secondary winding of 2 turns, then the turn ratio is:

   a. 6
   b. 4:2
   c. 2
   d. 1:2

17. The required branch-circuit conductor ampacity for X-ray equipment must be at least which of the following:

   a. 100 percent of the momentary demand rating of the two largest X-ray machines
   b. 100 percent of the momentary demand rating of the two largest X-ray machines plus 20 percent of the momentary demand of other X-ray units
   c. 100 percent of the continuous demand rating of the two largest X-ray apparatus
   d. 50 percent of the continuous demand rating of the two largest X-ray machines plus 10 percent of the momentary demand of other X-ray units

18. A 200 volt motor with a full-load current of 25 amperes requires a disconnect rated at which of the following:

   a. 15 amperes
   b. 25 amperes
   c. 28.75 amperes
   d. 50 amperes

19. A circuit passes through a ground rod with the resistance of 30 ohms and draws 5 amperes of current through the grounding conductor; at what voltage is the circuit operating?

   a. 350 volts
   b. 220 volts
   c. 208 volts
   d. 150 volts

20. If the secondary coil of a transformer produces 100 amperes at 240 volts, and the primary is energized at 480 volts, the transformer primary produces 120 amps.

   a. True
   b. False

21. A 120/240 three-wire feeder is installed in rigid PVC conduit running underground from a single-family dwelling unit to a detached workshop on the same property and is fed from a 90 amp CB in the dwelling panelboard; therefore, which of the following statements must be true:

    a. The feeder must terminate in a single disconnect.

    b. The feeder must have an overcurrent disconnect in the workshop.

    c. The feeder must have an equipment grounding conductor.

    d. The feeder must contain #8 conductors.

22. A section of stainless steel rod which is 8 feet long by 1/2 inch may serve as a grounding electrode.

    a. True   **b.** False

23. In a single-phase motor on an ungrounded, 2-wire, single-phase AC power source, the motor overload must be placed in which of the following conductors:

    a. Neutral conductor   **b.** Phase conductor

    c. Either of the above   **d.** None of the above

24. The type of load that increases heat in a transformer without operating its overcurrent device is which of the following:

    a. Nonlinear   **b.** Tri-phasic

    c. Continuous   **d.** Maximum allowable

25. If an 800 amp busway runs for 300 feet in a commercial building and the last 20 feet of the busway run is reduced to a rating of 200 amperes, then which of the following describes the installation of the smaller bus:

    a. It must be protected by an overcurrent device.

    b. It must be at least 1/3 the ampere rating of the larger bus.

    c. The total combined amperes must not exceed 240.

    d. All of the above

    e. None of the above

26. The highest voltage approved for a Class 1 power limited circuit is 30 volts.

    a. True   **b.** False

27. Meter readings on a three-phase electrical system that read 6000 watts on a wattmeter, 208 volts on a voltmeter and 20 amperes on an ammeter indicate a power factor of which of the following:

    a. .75 percent   **b.** .83 percent

    c. .92 percent   **d.** 1.44 percent

28. A receptacle connected to a 30 amp branch circuit shall have a maximum load of 24 amps and an ampere rating of which of the following:

    a. 12 amps   **b.** 24 amps

    c. 30 amps   **d.** None of the above

29. A building has two sets of service drop conductors from a utility pole where Drop A is 120/240 volt for lighting and receptacle loads and Drop B is a 480 volt motor load supply, which means that this building is required to have how many disconnecting means:

    a. 4   **b.** 8

    c. 9   **d.** 12

30. The minimum feeder load for a residential electric clothes dryer is 5000 watts using the standard calculation method.

    a. True   **b.** False

31. The minimum operating voltage for a 115V load connected to a 220V source is which of the following:

    a. 110 volts   **b.** 115 volts

    c. 123 volts   **d.** 209 volts

32. The total voltage across a series circuit is equal to the sum of the individual voltage drops.

    a. True   **b.** False

33. If there are more than three current-carrying conductors grouped together in a raceway then which of the following must occur:

    a. The cables must be separated by a nonmetallic barrier

    b. The ampacity must be derated

    c. The conductors shall not be run with electric light or power cables

    d. All of the above

34. The branch-circuit conductors for a fire pump motor must be sized so that the voltage at the line terminals of the controller, when the motor starts (locked-rotor current), does not drop more than 15 percent below the controller's rated voltage.

    a. True   **b.** False

35. The voltage drop of two 12 AWG THHN conductors that supply a 20A, 120V load located 100 feet from the power supply is which of the following:

    a. 6.4 volts   **b.** 7 volts

    c. 8 volts   **d.** 12 volts

36. When computing the load of a feeder, the load shall not be less than the sum of the loads on the branch circuits after demand factors, which may apply have been included.

    a. True
    b. False

37. Commercial buildings must have 15- and 20-ampere, 125-volt GFCI receptacles in which of the following locations:

    a. In the equipment room
    b. At grade level on the outside of the building
    c. In any bathrooms or kitchens
    d. All of the above

38. If the service entrance equipment for a single-family dwelling is rated 200 amps and is installed indoors, then it does not have to be illuminated.

    a. True
    b. False

39. Electrical metallic tubing (EMT) is marketed in which of the following configurations:

    a. Nonthreaded
    b. Threaded
    c. Plastic-coated corrosion resistant
    d. All of these

40. Each two feet of lighting track in a commercial occupancy is counted as which of the following when calculating service and feeder load requirements:

    a. 1440 watts
    b. 150 volt-amperes
    c. Two 20 amp circuits
    d. 110 volts

41. The metal standards supporting a raised floor in a computer room must be grounded.

    a. True
    b. False

42. The total power loss in a parallel circuit is equal to which of the following:

    a. The sum of the individual power dissipation
    b. The sum of the individual circuits multiplied by the length of the circuits
    c. The sum of the currents of the individual branches of the circuit
    d. None of the above

43. The maximum permitted horsepower for a size 1 NEMA rated 115 volt single-phase motor starter is which of the following:

- **a.** 1/3 horsepower
- **b.** 1/2 horsepower
- **c.** 1 horsepower
- **d.** 2 horsepower

44. Low-voltage Class 2 thermostat wire is permitted to be installed in the same raceway with power conductors for an air conditioning compressor.

- **a.** True
- **b.** False

45. In a 480-volt corner-grounded Delta system, the voltage-to-ground is which of the following:

- **a.** 480 volts
- **b.** 240 volts
- **c.** 220 volts
- **d.** 208 volts

46. Impedance in an alternating circuit results from which of the following:

- **a.** Capacitive reactance
- **b.** Inductive reactance
- **c.** Resistance
- **d.** All of the above

47. Which of the following shall be permitted to be used at any power takeoff opening to make power connections to a busway:

- **a.** Circuit-breaker cubicle
- **b.** Fusible switch adapter
- **c.** Cable tap box
- **d.** All of the above

48. The overcurrent protection device for a circuit that supplies a hermetic motor compressor must be rated for at least 225 percent or more of the rated current load under any circumstances.

- **a.** True
- **b.** False

49. Based on the figure below, the current in the neutral conductor is equal to which of the following:

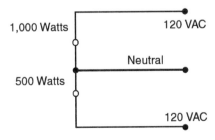

- **a.** 100 amperes
- **b.** 24.5 amperes
- **c.** 6 amperes
- **d.** 4.2 amperes

50. An equipotential plane is required for all indoor animal confinement areas with which of the following:

   a. More than 10 animals confined are housed in the space

   b. Areas housing birds or fowl only

   c. Portions of the area with a concrete floor

   d. More than 2000 sq feet

ANSWERS

TIMED TEST 2

1. **B** Section [250.53(F)]
2. **B** Section [210.19(A)(1)]
3. **A** Section [320.10(1)]
4. **A** Section [110.26]
5. **B** Section [250.118(7)(b)]
6. **A** Article 100
7. **A**

> ► **Test** tip
>
> The voltage drop will be in the same proportion as each resistor in a series is to the total circuit resistance. 28 ohms is 80 percent of the total resistance (35 ohms, which is 28 + 7) so 80 percent of the voltage will be across the 28 ohms resistor. 0.8×120 volts = 96 volts

8. **A** Article 430.102(B) Exception
9. **A** Section [220.40]
10. **D** Article 430.9(B)
11. **C** Section [430(B)(2)]
12. **B** Section [210.52(F)] Exception
13. **D** Section [517.30(B)(1)]
14. **A** Section [555.19(A)(4)]
15. **C** Section [517.19(A)]
16. **C** Basic Math
17. **B** Section [660.6(A)]
18. **C** 25 amperes $\times$ 115 percent = 25×1.15 = 28.75
19. **D** 30 Ohms $\times$ 5 Amperes = 150 volts

20. **B** $\dfrac{24000 \text{ (total of the volts} \times \text{amperes in the secondary)}}{480 \text{ (volts in the primary)}} = 50$

21. **C** Section [215.6]
22. **A** Section [250.52(A)(3)]
23. **C** Table [430.37]
24. **A** 450.3 FPN #2
25. **A** Section [368.17] Exception
26. **A** Section [725.21]
27. **B** PF = 6000 ÷ (208 × 20 × 1.73)
28. **C** Section [210.21]
29. **D** Section [230.71(A)]
30. **A** Section [250.24(5)] & [408.40]
31. **D**
32. **A**
33. **B**
34. **A**
35. **C**
36. **B** Article 310, Table [310.16]
37. **C**
38. **A** Section [230.24(B)(1)]
39. **A**
40. **B**
41. **B**
42. **A**
43. **D**
44. **B**
45. **A** Section [250.25(4)]
46. **D**
47. **D**
48. **A** Section [230.24(B)(1)]
49. **D**

> ► **Test** tip
> Current through 1000 watt load = 1,000/120 = 8.3 amperes; current through 500 watt load = 500/120 = 4.17 amperes. 8.3-4.1 = 4.2 amperes.

50. **C** Section [547.10]

—NOTES—

Chapter 15
TIMED TEST 3

Electrician's Exam Study Guide

Some states require longer tests than others. Almost all of the states only require that you score 70 percent to pass, and for the most part, those states that administer lengthier tests also give you more time to take them. In order to give you the best opportunity to practice your testing skills, we have provided you with a longer sample exam. This is still on open book test; however, this time you have four hours to take the exam. Before taking this test, go through your NEC® book and familiarize yourself once again with the locations of key subjects such as conduit, feeder and conductor sizing. Do you know off the top of your head what chapters to refer to for specialized systems, such as motor controls, hospital circuits, power limited systems, industrial machinery connections and classified locations? Do you already know that mobile home wiring is found in Article 550, and crane and hoist ratings and conductor sizes are in Article 610? Without having to flip to the Index of the NEC, you could memorize certain reference locations. For example, Table **[220.3]** lists the locations in the NEC® for load calculations, Table **[310.15(B)(2)(a)]** provides adjustment factors for raceways with more than three current-carrying conductors, and Tables **[310.16]** through **[310.21]** list the ampacities for various conductor types, temperature ratings and correction factors. Keep these and other key quick reference material locations, such as the tables in Chapter 9, in mind as you prepare to take this advanced test. Also remember the testing tips you have learned throughout this study guide. Now set your clock, get out your NEC® book and prepare to begin.

TIMED TEST # 3

TIME ALLOWED: 4 HOURS

1. If excessive amounts of metal in or on a building become energized and could be subject to personal contact, which of the following will provide additional safety measures:
   a. Adequate bonding and grounding
   b. Bonding
   c. A bonding jumper
   d. None of the above

2. The minimum size for copper lighting track conductors is which of the following:
   a. # 10 AWG
   b. #12 AWG
   c. #14 AWG
   d. #16 AWG

3. Type MV cable is rated for 2000 volts or less.
   a. True
   b. False

4. A fixture with a combustible material shade shall not be installed in locations where temperatures exceed which of the following:
   a. 30 degrees C
   b. 90 degrees C
   c. 90 degrees F
   d. 100 degrees F

5. Cable tray system shall not be used in hoist ways.
   a. True
   b. False

6. Conductors connected between the ground grid and all metal parts of a swimming pool must be which of the following sizes:
   a. #4
   b. #6
   c. #8
   d. #10

7. An outlet for specific appliances, including laundry equipment, must be located within how many feet of the appliance:
   a. 2 feet
   b. 4 feet
   c. 6 feet
   d. None of the above

8. The shortest distance measured between the top surface of the finished grade and the top surface of a direct buried conductor, cable, conduit or raceway is considered which of the following:

   a. Ground gap
   b. Direct-buried depth
   c. Soil compression
   d. Cover

9. Flexible metal conduit must be used to protect conductors that emerge from the ground and run to a pole.

   a. True
   b. False

10. In a metal raceway for recessed fixture connections, the tap connectors cannot exceed which of the following maximum lengths:

    a. 6 feet
    b. 4 feet
    c. 2 feet
    d. 1 foot

11. A restaurant with five pieces of kitchen equipment would have a feeder demand factor of which of the following:

    a. 50 percent
    b. 70 percent
    c. 75 percent
    d. None of the above

12. Conductors attached to a cablebus must be in the same raceway because of inductive current.

    a. True
    b. False

13. Based on the diagram below, the correct sequence for wiring a four-way switch is which of the following;

    To the 3-way switch

    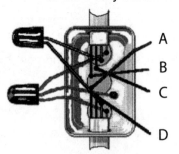

    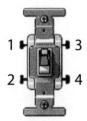

    To the 4-way switch

    4-way switch

    a. A to 3, B to 1, C to 3, and D to 2
    b. A to 2, B to 1, C to 3, and D to 4
    c. A to 1, B to 3, C to 2, and D to 4
    d. A to 1, B to 2, C TO 4, and D to 3

14. Branch circuits must have ground-fault circuit interrupters if the branch circuit supplies a swimming pool light fixture that exceeds which of the following:

    a. 10 volts   **b.** 15 volts

    c. 50 volts   **d.** 1 amp

15. Disregarding any exceptions, direct-buried cables must be buried at least 24 inches deep.

    a. True   **b.** False

16. Other than in dwelling units, any 15 amperes and 20 amperes receptacles must have a ground-fault circuit interrupter protection in which of the following locations:

    a. Bathrooms   **b.** Rooftops

    c. Kitchens   **d.** All of the above

17. Eight #6 XHHW conductors running in schedule 40 PVC conduit requires which of the following sizes of rigid conduit:

    a. 1 1/4 inches   **b.** 1 inch

    c. 3/14 inch   **d.** 1/2 inch

18. One-inch rigid conduit with three #10 TW conductors must have a minimum radius bend of which of the following:

    a. 5 3/4 inches with a one-shot bender

    b. 6 inches with a one-shot bender

    c. 10 inches with a one-shot bender

    d. None of the above

19. Metal enclosures used to protect cable assemblies from physical damage are not required to be grounded.

    a. True   **b.** False

20. Which of the following amperes is not a standard classification for a branch circuit that supplies several loads:

    a. 50 amperes   **b.** 30 amperes

    c. 25 amperes   **d.** 20 amperes

21. Group operated switches shall be used for capacitor switching.

    a. True   **b.** False

22. Which of the following are required for temporary wiring:
    a. All branch circuits must originate in an approved panelboard
    b. Flexible cords must be protected from accidental damage
    c. All conductors must have overcurrent protection

23. A 120-gallon electric water heater shall have a branch circuit rating not less than which of the following:
    a. 80 percent of the water heater rating
    b. 100 percent of the water heater rating, plus 5 percent of the overall circuit rating
    c. 115 percent of the water heater rating
    d. 125 percent of the water heater rating

24. Conductors size No. 1 that are run for general use may not be run in parallel.
    a. True
    b. False

25. A minimum of 4 inches of free conductor must remain at each outlet in order to connect luminaires.
    a. True
    b. False

26. Upon completion of construction or the conclusion of the purpose for which it was installed, temporary wiring shall be removed within which of the following time frames:
    a. 31 days after an event
    b. 60 days after the conclusion of an event or construction
    c. 90 days after the completion
    d. Immediately

27. A traveling cable shall be allowed to run without using a raceway under which of the following conditions:
    a. The distance from the first point support does not exceed 6 feet.
    b. The conductors are grouped and taped together.
    c. The conductors are in their original sheath.
    d. All of the above

28. Conductors that supply a fire pump motor must have a rating not less than ___ percent of the sum of the fire pump motor's full load current and ___ percent of any associated fire pump accessory equipment:
    a. 80, 100
    b. 125, 100
    c. 115, 125
    d. 100, 100

29. The smallest fixture wire permitted for general purpose installations shall be #18:

    a. True   **b.** False

30. A continuous white or natural gray covering can only be used on which of the following types of conductors:

    a. Grounding   **b.** Neutral

    c. Grounded   **d.** Hot

31. Equipment connected to a 40 amp circuit requires that the minimum size of the copper equipment grounding conductor be no less than which of the following:

    a. #8   **b.** #10

    c. #12   **d.** #14

32. A #14 grounded conductor in a circuit must be either a white or gray color.

    a. True   **b.** False

33. The antenna for a satellite system must be grounded to which of the following

    a. Interior metal water piping system, within 5 feet from its point of entrance

    b. The building or structure grounding electrode system

    c. An accessible means external to the building

    d. All of the above

34. Rigid 2-inch metal conduit shall be supported at a distance no less than which of the following:

    a. Every 8 feet   **b.** Every 12 feet

    c. Every 14 feet   **d.** Every 16 feet

35. When 6 current-carrying conductors are installed in a raceway, the maximum load of each must be reduced by 80 percent.

    a. True   **b.** False

36. Type UF cable shall be used for which of the following applications:

    a. Concrete encased   **b.** Direct buried

    c. Service entrance cable   **d.** None of the above

37. Service disconnect means shall clearly indicate which of the following:

   a. The minimum and maximum voltage rating

   b. If it is in the open or closed position

   c. Both of the above

   d. None of the above

38. If a nipple contains six current-carrying #6 THW copper conductors, then the ampacity of each conductor would be which of the following:

   a. 32.5 amps
   b. 40 amps
   c. 65 amps
   d. 110 amps

39. Type AC cable shall not be used to connect a power motor to a motor supply.

   a. True
   b. False

40. A 3-phase AC unit must be directly connected to a recognized wiring method and may not include a cord connection.

   a. True
   b. False

41. Bathroom receptacle outlets must be supplied by which of the following:

   a. Ground fault protection

   b. At least one 30 amp branch circuit

   c. Both of the above

   d. None of the above

42. The total connected load in watts of three electric heaters rated 1000 watts and 1100 watts at 120 volts and 800 watts at 240 volts is:

   a. 2417 watts
   b. 2900 watts
   c. 3100 watts
   d. 5800 watts

43. An electric toaster oven rated at 1440 watts plugged into a 120 volt receptacle in a single-family dwelling will draw 16 amperes.

   a. True
   b. False

44. MC cable may not be installed in a cable tray.

   a. True
   b. False

45. The voltage across the 5 amp load in the figure below, once the neutral is opened, must be which of the following:

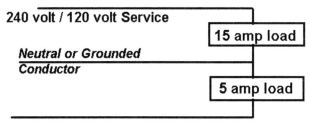

-Voltage across Resistance Load-

- **a.** 120 volts
- **b.** 180 volts
- **c.** 240 volts
- **d.** 720 volts

46. A 25 horsepower, 208V three-phase fire pump motor circuit overcurrent protective device must be sized in accordance with which of the following:

- **a.** To carry indefinitely the sum of the locked-rotor current of the fire pump
- **b.** Be 450 amps
- **c.** Both of the above
- **d.** None of the above

47. The nine conductors as follows-four #12 copper THWN, three #6 RHW covered copper, and two #8 covered copper RHW-can be installed in rigid metal conduit that is no smaller than which of the following:

- **a.** 25 feet long
- **b.** 5325 square inches
- **c.** 1 1/4 inches
- **d.** All of the above

48. The demand factor percent for the portion of a lighting load greater than 120,000 VA is which of the following:

- **a.** 35 percent
- **b.** 50 percent
- **c.** 100 percent
- **d.** None of the above

49. The current flow for a 120 volt outlet that has an electric hair dryer with a resistance of 15 ohms plugged into it is which of the following:

- **a.** 8 amperes
- **b.** 12 amperes
- **c.** 15 amperes
- **d.** 30 amperes

50. A 240 volt, 3-phase electric heater that draws 11 amperes and has a power factor of 1.0 produces heat equal to which of the following:
   a. 2640 watts of heat
   b. 2200 watts of heat
   c. 4567 watts of heat
   d. 5280 watts of heat

51. Total resistance of a series circuit is equal to which of the following:
   a. The sum of each individual resistance multiplied by the number of circuits in the series
   b. The sum of the individual resistances
   c. The number of circuits in the series divided by the sum of the individual resistances
   d. None of the above

52. A terminal may be connected to a #8 conductor with either a pressure connector or a solder lug.
   a. True
   b. False

53. A 480 volt corner-grounded Delta system has a voltage-to-ground of which of the following:
   a. 480 volts
   b. 240 volts
   c. 120 volts
   d. 2400 amperes

54. A portable generator that draws 746 watts operates at which of the following:
   a. Maximum efficiency
   b. 1 horsepower
   c. A load requiring a 50 amp circuit
   d. None of the above

55. A 240 volt multi-wire branch circuit that has a branch circuit protection device that opens each ungrounded conductor at the same time may be used to supply line-to-line loads.
   a. True
   b. False

56. A continuous load is one in which the maximum current is expected to continue for at least which of the following:
   a. 24 hours, non-stop
   b. 12 hours or more
   c. 6 hours or more
   d. 3 hours or more

57. The diagram below is an example of which of the following:

a. 3-wire "Y"
b. 4-wire Delta
c. 3 wire ground to disconnect
d. None of the above

58. When installing rigid metal conduit (RMC), there cannot be more than the equivalent of four quarter bends (360 degrees) between which of the following:

a. Conduit bodies
b. Outlet boxes
c. Junction boxes
d. All of the above

59. A stationary 1/2 horsepower motor may have a branch circuit overcurrent device as its disconnecting means.

a. True
b. False

60. A nonmetallic wireway is not permitted in which of the following locations unless it is specifically marked otherwise:

a. Where the wireway would be exposed to sunlight
b. Where subject to corrosive vapors
c. In wet locations
d. All of the above

61. The dimensions of a pull box for a straight run of #4/3 Romex must meet which of the following size requirements:

a. The length of the box cannot be less than 8 times the trade diameter of the raceway.
b. The overall dimensions of the box cannot exceed 2 square feet.
c. The width of the box cannot be less than 6 times the trade diameter of the raceway.
d. All of the above

62. A fire-resistant room or enclosure must surround a transformer with a rating greater than which of the following:

a. 112 1//2 kVA
b. 120 kVA
c. 132 1/2 kVA
d. 25,000 volts

63. Feeders that contain a common neutral are permitted to supply which of the following:

    a. One, two or three sets of 3-wire feeders

    b. Two sets of 4-wire feeders

    c. Two sets of 5-wire feeders

    d. All of the above

64. The element of an electrical system that is intended to carry but not to utilize electric energy would be which of the following:

    I. Snap switch   II. Receptacle   III. Device   IV. Light bulb

    a. I only   **b.** IV only

    c. I and II only   **d.** I, II and III

65. The minimum size for a 10,000 volt copper conductor is which of the following:

    a. #10 AWG   **b.** #6 AWG

    c. #2 AWG   **d.** 1/0 AWG

66. A circuit protected by a 50 ampere circuit breaker requires which of the following copper equipment grounding conductors:

    a. #14 AWG copper   **b.** #12 AWG copper

    c. #10 AWG copper   **d.** #8 AWG copper

67. When an EMT raceway contains 12 ten-feet long #12 THHN copper conductors where there are nine current carrying conductors and the ambient temperature is 42 degrees C, then the ampacity of the THHN conductors is closest to which of the following ratings:

    a. 15 amperes   **b.** 18 amperes

    c. 25 amperes   **d.** 30 amperes

68. Disregarding any exceptions, a conductor rated 56 amperes shall be protected by a fuse sized at which of the following:

    a. 60 amp   **b.** 50 amp

    c. 30 amp   **d.** 20 amp

69. A luminaire may not be supported by the screw shell of a lampholder if it exceeds which of the following:

    a. Weighs more than 6 pounds   **b.** 16 inches in any dimension

    c. Both of the above   **d.** None of the above

70. Light and power systems conductors 600 volts or less may occupy the same enclosure only when all of the conductors meet which of the following requirements:

    a. All of the conductors are insulated.

    b. All conductors are insulated to meet the maximum temperature rating for the enclosure.

    c. All conductors are insulated for the maximum voltage applied to any conductor in the enclosure.

    d. None of the above

71. Disregarding any exceptions, a copper conductor used for overhead services must be at least which of the following sizes:

    a. #8   **b.** #10

    c. #12   **d.** #14

72. Nine #8 AWG THHN copper conductors in a conduit at 30 degrees C have a maximum allowable ampacity of which of the following:

    a. 33 amperes each   **b.** 38.5 amperes each

    c. 55 amperes each   **d.** 66 amperes each

73. The minimum individual amperes for two hot, ungrounded feeder circuit conductors running to a commercial building that has a general lighting load of 50,000 volt amperes, supplied by a single-phase, 240 volt feeder circuit is closest to which of the following:

    a. 208 amperes   **b.** 217.39 amperes

    c. 228 amperes   **d.** 480 amperes

74. A 25 feet long feeder tap must have an ampacity of the feeder conductor or overcurrent protection device from which it is tapped that is at least which of the following:

    a. .25   **b.** 1/3

    c. .5   **d.** 3/4

75. Disregarding any exceptions, a copper-clad aluminum conductor used for overhead services must be at least which of the following sizes:

    a. #12   **b.** #10

    c. #8   **d.** #6

76. Based on the figure below, assuming the box is used for splices and does not have the cubic inch dimensions marked on it and No. 4 or larger conductors are installed in the raceways, then the dimension for "X" shall be which of the following:

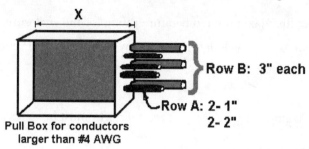

Pull Box for conductors larger than #4 AWG

- **a.** 12 inches
- **b.** 16 inches
- **c.** 18 inches
- **d.** 24 inches

77. The power factor for a 6500-watt load connected to a 240 volt, AC, single-phase 36 amp circuit is which of the following:

- **a.** .75 or 75 percent
- **b.** .8 or 80 percent
- **c.** .93 or 93 percent
- **d.** None of the above

78. Tap conductors must terminate in a single circuit breaker or single set of fuses, which will limit the load to the ampacity of the tap conductors.

- **a.** True
- **b.** False

79. Based on the illustration below, Voltage A should be which of the following.

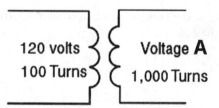

- **a.** 120 volts
- **b.** 240 volts
- **c.** 1200 volts
- **d.** 12,000 volts

80. A multi-wire branch circuit that supplies more than one device on the same yoke in a dwelling unit must be protected by which of the following:

- **a.** One fuse
- **b.** Two single-pole circuit breakers that are tied together
- **c.** Two single-pole circuit breakers that operate independently
- **d.** One single-pole non-fused disconnect

81. Which of the following receptacles may be connected to small appliance circuits:

   a. A garage ceiling receptacle for a garage door opener

   b. A living room receptacle for a television set

   c. A receptacle in the kitchen for an electric clock

   d. None of the above

82. Conductors are considered exposed if they are installed 7 feet or less from the top of the floor.

   a. True   **b.** False

83. Unless a recessed light fixture is identified or marked as being suitable for direct contact with insulation, thermal insulation cannot be installed within which of the following distances from a recessed light fixture:

   a. 1 1/4 inches   **b.** 3 inches

   c. 6 inches   **d.** 8 inches

84. Three resistors, rated 5, 10, and 15 ohms respectively, are connected in parallel, which produces a combined total resistance of which of the following:

   a. 2.7 ohms   **b.** 3.33 ohms

   c. 18.33 ohms   **d.** 30 ohms

85. A circuit conductor that is intentionally grounded is which of the following:

   a. Grounded conductor

   b. Equipment grounding conductor

   c. Grounding conductor

   d. Grounding electrode

86. The maximum total voltage drop of two feeders and three branch circuits to the farthest outlet in a wiring system should not exceed which of the following:

   a. 2 percent for the feeders

   b. 3 percent for the branch circuits

   c. 5 percent

   d. 10 percent maximum

87. The maximum size permitted for flexible metallic tubing installed in a commercial office building would be which of the following:

   a. 1 inch   **b.** 3/4 inch

   c. 1/2 inch   **d.** 1/4 inch

88. Open wiring on insulators installed 6 feet from floor level do not require protection.
   a. True
   b. False

89. Class II locations are considered hazardous because which of the following is present in quantities sufficient to create the potential for combustion:
   a. Combustible dust
   b. Ignitable vapors
   c. Combustible flyings
   d. None of the above

90. Each kitchen in a two-family dwelling unit requires a minimum of which of the following in order to meet the small appliance load:
   a. Two 15 amp circuits
   b. One 20 amp circuit
   c. Two 20 amp circuits
   d. Four 20 amp circuits

91. Based on the figure below, the total resistance in ohms would be which of the following:

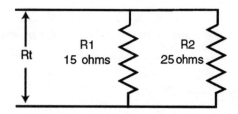

   a. 3.75
   b. 4.0
   c. 9.4
   d. 40

92. A 12-kilowatt electric range in a dwelling unit must have an 8-kilowat branch circuit load.
   a. True
   b. False

93. Flexible cords are not allowed for use in which of the following locations:
   a. To replace fixed wiring of a structure
   b. Hoists
   c. Cranes
   d. All of the above

94. The inside of a paint spray booth is considered which of the following locations:
   a. Class 1, Division 1
   b. Class 1 or Class 2, Division 1
   c. Class 3, Division 1
   d. None of the above

95. Grounding electrodes perform which of the following functions:

   a. Prevent current flow into the earth

   b. Absorb stray currents

   c. Return neutral current to the earth

   d. Make an electrical connection to the earth

96. The function of a supplementary overcurrent protective device is best described by which of the following statements:

   a. Is not a substitute for branch circuit overcurrent protection

   b. Is required to be installed in light fixtures

   c. Must be installed in any branch circuit system

   d. Shall not required on circuits rated more than 50 amps

97. Which of the following statements is considered mandatory in the NEC:

   a. See UL Standard 1699 on Arc-Fault Circuit Interrupters

   b. [NFPA® 70e]

   c. FPN: For conductors exposed to deteriorating agents, see 310.9

   d. Circuits and equipment shall comply with 820.3(A) through 820.3(G)

98. The maximum allowable voltage for a Class 1 power-limited circuit is which of the following:

   a. 96 volts   **b.** 60 volts

   c. 30 volts   **d.** 20 volts

99. Two resistors connected in a series across a 120 volt power source, where the first has a resistance of 28 ohms and the second has a resistance of 7 ohms, will generate a voltage drop across the first 28 ohms resistor of which of the following:

   a. 96 volts   **b.** 60 volts

   c. 28 volts   **d.** 120 volts

100. Direct-buried conductors and cables that emerge from the ground to a pole should be protected in flexible metal conduit.

   a. True   **b.** False

ANSWERS

TIMED TEST 3

1. **A** Section [250.116] FPN
2. **B** Section [410.105(A)]
3. **B** Section [328.2]
4. **B** Section [410.5]
5. **A** Section [392.4]
6. **C** Section [680.26(C)]
7. **C** Section [210.50(C)]
8. **D** Table [300.5]
9. **B** Section [438.12(6)]
10. **A** Section [410.67(C)]
11. **B** Table [220.55]
12. **A** Section [300.20(A)]
13. **B**
14. **B** Section [680.23(A3)]
15. **A** Section [300.5(A)] and Table [300.5]
16. **D** Section [210.8(B)(1)–(3)]
17. **A** Table 4
18. **A** Table [344.24] & Chapter 9, Table 2
19. **A** Section [250.86] Exception 2
20. **C** Section [210.3]
21. **A** Section [460.24(A)]
22. **D** Section [590.4(C) & (H)]
23. **D** Section [422.13]
24. **A** Section [310.4]
25. **B** Section [300.14]
26. **D** Section [590.3(D)]
27. **D** Section [620.44]
28. **B** Section [695.6(C)(1)]
29. **A** Section [402.3] and Table [402.3]
30. **C** Section [200.7]
31. **B** Table [250.122]
32. **B** Section [200.6(A)]
33. **D** Sections [810.15]
34. **D** Table [344.30(B)(2)]

35. **A** Section [310.15(B)(2)(a)]
36. **B** Section [340.10(1)]
37. **B** Section [230.77]
38. **C** Section [310.15(B)(2)] Exception 3
39. **B** Sections [320.10] & [320.12]
40. **A** Section [440.60]
41. **A** Section [210.8(A)(1)] & [210.52(D)]
42. **A** Add the wattage of each heater. Voltage is not a factor.
43. **B** Amperes = $\dfrac{\text{power (watts)}}{\text{volts}}$ so $\dfrac{1440 \text{ watts}}{120 \text{ volts}} = 12$ Amperes
44. **B** Table [392.3(A)]
45. **B**

> ► **Test** tip
>
> First find the resistance of each load:
>
> 15 ampere load: $R = E/I$; Volts = 120; $R = 120$ volts/15 amperes; $R = 8$ ohms
>
> 5 ampere load:
>
> $R = 120$ volts/5 amperes; $R = 24$ ohms
>
> When neutral is open, the two resistors are in series and the Total Resistance is $8 + 24 = 32$ ohms.
>
> The voltage is 240 volts across the two resistors in series.
>
> To find the current after the neutral opens:
>
> $I = 240$ volts/32 ohms; $I = 7.5$ amperes
>
> The voltage across the 5 ampere load that has a resistance of 24 ohms is:
>
> $E = I \times R$; $E = 7.5$ ampere $\times$ 24 ohms = 180

46. **C** Section [695.4(B)(1)]. Also, according to Table [430.251(B)], the locked-rotor current of a 25 hp, 208V, three-phase motor is 404A, and Section [240.6] requires a 450A protection device.
47. **C**

> ► **Test** tip
>
> From Chapter 9, Table 5:
>   Area of #12 THWN = .0133 × 4 = .0532
> Area of # 8 RHW covered = .0835 × 2 = .167
> Area of #6 RHW covered = .1041 × 3 = .3123
>   Total area = .5325 square inches
> See Table 4 for Rigid Metal Conduit. Find the first number larger than .5325, which is listed as 0.610 in the column for "Over 2 Wires - 40%", which is 1 1/4 inch conduit

48. **D** Section [220.11]
49. **A** Amperes = $\dfrac{\text{volts}}{\text{resistance}}$ so $\dfrac{120 \text{ volts}}{15 \text{ ohms}}$ = 8 Amperes
50. **C** Power = $\sqrt{n}$ × Volts x Amperes × power factor
    1.73 × 240(volts) × 11 amperes) × 1.0 = 4567 Watts
51. **B**
52. **A** Section [110.14(A)]
53. **A** Section [250.26(4)]
54. **B**
55. **A** Section [210.4(C)] Exception #2
56. **D** Article 100
57. **B**
58. **D** Section [344.26]
59. **B** Section [430.109(B)]
60. **A** Section [378.12(3)]
61. **A** [314.28(A)(1)]
62. **A** Section [450.21(b)]
63. **D** Section [215.4(A)]
64. **D** Article 100 Definitions
65. **C** Table [310.5]
66. **C** Table [250.122]
67. **B** Section [310.15(B)], Table [310.16], and Correcting Factors Table [310.15(B)(2)(a)]. NOTE: The derating factor for 9 current carrying conductors in a raceway is 70 percent (.70), and the ambient temperature derating factor for 42 degrees C is .87. So even though the ampacity of #12 THHN is 30 amps, when you figure it including the derating factors:

    30 amps × .7 × 87 = 18.27 amps

68. **A** Sections [240.4(B)] and [240.6(A)]
69. **C** Section [410.15(A)]
70. **C** Section [300.3(C)(1)]

> ▶ **Test** tip
> Remember that you must choose the most current answer. Answer A is not as comprehensive an answer as C.

71. **A** Section [230.33(B)]
72. **B** Section [310.16] and Table [310.15(B)(2)(a)]
    NOTE: Derating factor for the 9 conductors in a raceway is 70 percent, or .7 × 55 (the factor for 30 degrees C), which equals 38.5.

73. **A** Appendix D

> ► **Test** tip
>
> Ohm's law $I = P \div E$. This equates to $50{,}000 \div 240$ volts $= 208$. Additionally the question is looking for individual amperes for the conductors; it does not matter how many of them are listed in the question.

74. **B** Section [241.21(B)(2)(1)]
75. **D** Section [230.33(B)]
76. **D** Section [314.28(A)(2)]
    NOTE: You must use the larger of either Row A $(6 \times 2) + 2 + 2 + 1 + 1 = 16$, or Row B, which is $(6 \times 3) + 3 + 3 = 24$ inches, so the dimension for "X" has to be 24 inches.
77. **A** General Electrical Knowledge Calculation: Power Factor is the volts multiplied by the amps, and then the watts are divided by that number (240 volts $\times$ 36 amps) $= 8640$; 6500 watts $\div 8640 = 0.75$
78. **A** Section [241.21(B)(2)(1)]
79. **C** General Electrical Knowledge Calculation: $120 \div$ Voltage $= 100 \div 1000$; $120 \times 1000 = 120000.00 \div 100 = 1200$ volts
80. **B** Section [210.4(B)]
81. **C** Section [210.52(B)(2)] Exception #1
82. **A** Section [398.15]
83. **B** Section [410.66(B)]
84. **A** General Electrical Knowledge Calculation: $1/RT = 1 \div (1/RT + 1/R3)$ so $5 \times 10 = 50$ and $5 + 10 = 15$; $50 \div 15 = 3.33$; $15 \times 3.33 = 49.95$; $15 + 3.33 = 18.33$; $49.95 \div 18.33 = 2.73$ ohms
85. **A** Article 100
86. **C** Section [210.19(A)(1)] and Section [215.2(A)(4)] FPN #4

> ► **Test** tip
>
> This is a good example of why it pays to read the question carefully. The question asks for "total maximum voltage drop," so the number of feeders or branch circuits is irrelevant. All that matters is that the code requires 2 percent voltage drop for feeders and 3 percent voltage drop for branch circuits, and $2 + 3 = 5$ for a total.

87. **B** Section [360.20(B)] NOTE: The maximum size requirement is not related to the construction type, so it does not matter if this is a commercial office building.
88. **B** Section [398.15]
89. **A** Section [500.6(B)]

90. **C** Section [210.11(C)(1)]
91. **C** General Electrical Knowledge Calculation: Product over Sum 15 × 25 = 375; 15 + 25 = 40; 375 ÷ 40 = 9.38. Answer C is the closest to this sum.
92. **A** Section [220.55] and Table [220.55]
93. **A** Sections [400.7] and [400.8]
94. **B** Table [516.3(B)]
95. **D** Article 100
96. **A** Article 100
97. **D** Article [90.5(C)]
98. **C** Section [725.21]
99. **A** General Electrical Knowledge: The voltage drop will be in the same proportion as each resistor in a series is to the total circuit resistance. 28 ohms is 80 percent of the total resistance (35 ohms, which is 28 + 7), so 80 percent of the voltage will be across the 28 ohms resistor. **0.8 × 120 volts = 96 volts**
100. **B** Section [348.12(6)]

Chapter 16
CLOSED BOOK EXAM

Electrician's Exam Study Guide

The following practice exam is designed as a closed-book test. You will have 4 hours to complete 100 questions. In order to get a passing grade, you must get 70 questions correct without referring to the NEC® book for any assistance.

1. Both the size and setting for a circuit breaker used as an overcurrent protection device are the same.
   a. True
   b. False

2. A separate receptacle load is not required in guest rooms in hotels and motels.
   a. True
   b. False

3. A factor that would affect conductor ampacity is which of the following:
   a. The conductor length
   b. Voltage
   c. Temperature
   d. Motor size

4. A protective device used for limiting surge voltages by discharging surge current and that prevents continued flow of follow current while still remaining capable of repeating these functions is considered which of the following:
   a. Circuit breaker
   b. Fuse
   c. Disconnect switch
   d. Surge arrestor

5. Rod and pipe grounding electrodes must not be less than which of the following lengths:
   a. 5 feet
   b. 6 feet
   c. 8 feet
   d. 10 feet

6. Alternating current may be increased or reduced by using which of the following:
   a. By-pass switch
   b. A transformer
   c. Grounding conductor
   d. None of the above

7. Which of the following conductors has one or more layers of non-conducting materials that are not considered insulation:
   a. Covered
   b. Wrapped
   c. Rubber-coated
   d. All of the above

Closed Book Exam

8. Conductor sizes are listed as which of the following:
   - a. Diameter
   - b. Area
   - c. AWG or circular mils
   - d. AWG or millimeters

9. If a circuit voltage is increased and all other factors remain the same, then which of the following will change:
   - a. Current
   - b. Resistance
   - c. Ampacity
   - d. All of the above

10. The primary purpose for using thermal overload relays for polyphase induction motors is to protect against which of the following:
    - a. Fire
    - b. Short circuits between phases
    - c. Low voltage
    - d. Sustained overload

11. A multi-wire branch circuit protected by fuses may supply which of the following:
    - a. Only one load
    - b. Only line-to-neutral loads
    - c. Both of the above
    - d. Neither of the above

12. The minimum headroom at a motor control center must be 6 feet.
    - a. True
    - b. False

13. The ratio of watts to volt-amperes in a DC circuit is which of the following:
    - a. Unknown
    - b. Unity
    - c. Greater than one
    - d. 2:1

14. To connect rigid metal conduit, which of the following wrench types would not be used:
    - a. Stillson
    - b. Chain
    - c. Strap
    - d. Box end

15. In an industrial building, branch circuits protected by 15 amp overcurrent devices are permitted to supply a power load that operates at 300 volts or less.
    - a. True
    - b. False

16. Excluding those in the mains, the maximum number of overcurrent devices permitted in a single lighting and appliance panelboard is which of the following:

- **a.** 12
- **b.** 24
- **c.** 36
- **d.** 42

17. Rigid nonmetallic conduit shall be permitted to have a total number of quarter bends in one run that does not exceed which of the following:

- **a.** 2
- **b.** 3
- **c.** 4
- **d.** None of the above

18. The horsepower rating of a motor is which of the following:

- **a.** The motor speed
- **b.** The output of the motor
- **c.** Cannot be converted to watts
- **d.** The required input of the motor

19. Which of the following shall be used to connect the grounding terminal of a grounding receptacle to a grounded box:

- **a.** A junction box
- **b.** A main bonding jumper
- **c.** A neutral conductor
- **d.** An equipment bonding jumper

20. The residual voltage of a capacitor must be reduced to which of the following voltage levels within one minute after the capacitor is disconnected from the power source:

- **a.** 50 volts or less
- **b.** 60 volts or less
- **c.** 120 volts or less
- **d.** Zero volts

21. If an installation has two 2-wire branch circuits then the service disconnecting means must be rated at not less than which of the following:

- **a.** 15 amps
- **b.** 30 amps
- **c.** 50 volts
- **d.** 110 volts

22. The required distance between outlets along the floor line of any wall space in the living room of a dwelling unit is which of the following:

- **a.** No more than 6 feet measured horizontally from one outlet to another
- **b.** Every 10 feet
- **c.** Only one outlet is required per wall
- **d.** None of the above

23. In a dwelling unit, a 40 amp branch circuit is permitted to supply several fixed electric heaters:

    a. True   **b.** False

24. Live parts on equipment operating at 120 volts must have a working clearance in front of them of at least three feet:

    a. True   **b.** False

25. An auxiliary gutter is not permitted to extend for more than which of the following distances:

    a. 6 feet   **b.** 10 feet

    c. 20 feet   **d.** None of the above

26. Single phase loads connected on the load side of a phase converter shall be permitted to be connected to which of the following:

    a. Grounded phase   **b.** Neutral

    c. High leg   **d.** All of the above

27. The ampacity of a conductor must be derated where the ambient temperature exceeds which of the following:

    a. 150 degrees F   **b.** 112 degrees F

    c. 86 degrees F   **d.** 20 degrees C

28. The type of switch required in each location to independently control a light fixture from two different locations is which of the following:

    a. Single pole, double throw   **b.** Double pole, single throw

    c. Double pole, double throw   **d.** Three-way single throw

29. Which of the following means is not used to fasten equipment to concrete:

    a. Lead shield   **b.** Steel bushing

    c. Expansion bolts   **d.** All of the above

30. A 200 amp panel is permitted to supply a 200 amp noncontinuous load:

    a. True   **b.** False

31. Under no conditions shall a receptacle be permitted to be installed within which of the following distances from the inside wall of a pool:

    a. 3 feet   **b.** 6 feet

    c. 10 feet   **d.** All of the above

32. Heavy duty lamps are used on which of the following circuit sizes:
    a. 25 amps or larger
    b. 20 amps or larger
    c. 15 amps or larger
    d. None of the above

33. Low torque, low speed synchronous motors used to drive such equipment as reciprocating compressors or pumps that start unloaded do not require a fuse rating greater than which of the following:
    a. 100 percent of the continuous load
    b. 150 percent of the volt-amperes
    c. 200 percent of the full load current
    d. 185 percent of the run voltage

34. The definition of a pool includes wading, swimming and therapeutic pools, and the term *fountain* includes which of the following:
    I. Reflection pools
    II. Drinking fountain
    III. Display pool
    IV. Ornamental pool

    a. II only
    b. IV only
    c. I, III & IV
    d. All of the above

35. A light fixture installed under a canopy in front of a delicatessen is considered to be in which of the following locations:
    a. Commercial Class II
    b. Dry
    c. Wet
    d. Damp

36. The decimal equivalent of the fraction 9/16 is which of the following:
    a. 0.675
    b. 0.7625
    c. 0.837
    d. None of the above

37. Formal interpretation of the National Electrical Code is found in which of the following:
    a. NFPA® 70E
    b. International Electrical Code
    c. NFPA® Regulations Governing Committee Project
    d. OSHA Standards

38. Solid dielectric insulated conductors in permanent installations operating over 2000 volts shall meet which of the following requirements:
    a. Be shielded
    b. Have ozone-resistant insulation
    c. Both A & B
    d. Neither of the above

39. An individual branch circuit shall supply only one piece of utilization equipment.

   a. True   **b.** False

40. The minimum service load for each individual mobile home in a mobile home park is which of the following:

   a. 12,000 volt-amperes

   b. The greater of 14,000 volt-amperes or the calculated load

   c. 16,000 volt-amperes

   d. None of the above

41. The maximum allowable voltage drop for branch circuits and feeders combined shall not exceed 3 percent of the circuit voltage.

   a. True   **b.** False

42. A lamp that is 1000 watt, 120-volt will use the same amount of energy as a 14.4 ohm resistor at which of the following:

   a. 1200 watts   **b.** 240 volts

   c. 144 volts   **d.** 120 volts

43. The separation distance from non-current-carrying metal parts of electrical equipment and lightning protection conductors is which of the following:

   a. 6 feet through air

   b. 3 feet through dense materials

   c. 3 feet through concrete

   d. All of the above

44. Transformers rated over 35,000 volts must be installed in vaults:

   a. True   **b.** False

45. In electrical construction, mica is most commonly used for which of the following:

   a. In switchboard panels   **b.** Communicator bus separators

   c. Conductor insulators   **d.** None of the above

46. Edges that are invisible in an electrical drawing should be represented by which of the following line type:

   a. Solid   **b.** Dotted

   c. Broken   **d.** All of the above

47. Taps and splices shall not be located within which of the following fixture components:

   a. Stems and arms
   b. Splice boxes
   c. Ring assembly
   d. None of the above

48. Floor boxes have to meet the same requirements for spacing as wall receptacles if the box is which of the following:

   a. 24 inches from the wall
   b. Within 18 inches of the wall
   c. Within a 12-inch radius of the wall
   d. Floor boxes never have to meet wall receptacle requirements.

49. When computing ampacity, a fraction of an ampere is considered only when it is which of the following:

   a. Over 75 percent
   b. 0.5 or larger
   c. 0.25 to 0.5
   d. Never

50. If a single-family dwelling with 2050 square feet of living space includes a kitchen, two bedrooms, a living room, dining room, and a bathroom with 7 feet of counter space including the sink, then the total number of receptacles required to serve the bathroom area is which of the following:

   a. 4
   b. 3
   c. 2
   d. 1

51. When installed in a ventilated channel cable tray with a 4-inch inside width, the sum of the diameters of all of the single conductors shall meet which of the following demands:

   a. Shall be less than the width of the channel
   b. Shall be permitted to total 8 inches or less
   c. Shall not exceed 4 inches
   d. None of the above

52. The minimum electrical requirements for a dwelling to meet code is which of the following:

   I. Two small appliance circuits
   II. One laundry circuit
   III. 3 volt-amps per square foot
   IV. One 8 kw range connection

   a. I & II
   b. III only
   c. I, II & III
   d. All of the above

53. Which of the following is permitted to be connected ahead of service switches:

    a. Current-limiting devices   **b.** Surge arrestors

    c. Both of the above   **d.** None of the above

54. If an enclosure is supported by a suspended ceiling system, then it shall be fastened to the framing by which of the following methods:

    a. Screws   **b.** Rivets

    c. Bolts   **d.** All of the above

55. Which of the following shall not be used in wet or damp locations:

    a. Open wiring   **b.** AC armored cable

    c. Both of the above   **d.** None of the above

56. Removing which of the following from conduit threads will ensure electrical continuity between conductors:

    a. Enamel   **b.** Copper ends

    c. Rubber coating   **d.** All of the above

57. A panel may have from 2 to 6 disconnecting means, but which of the disconnect means listed below is permitted to be remote from the others:

    a. Elevator   **b.** Hoist

    c. Water pump for fire protection   **d.** Nurse Call system

58. A freestanding office partition is permitted a maximum of how many 15 amp receptacles:

    a. 15   **b.** 13

    c. 12   **d.** 10

59. In a commercial building, which of the following locations must have GFCI protection:

    a. The hallway   **b.** Outdoor receptacles

    c. Bathroom receptacles   **d.** All of the above

60. If a transformer vault is not protected by an automatic sprinkler system, then it must have a minimum fire resistance and structural strength of which of the following periods of time:

    a. 2 hours   **b.** 3 hours

    c. 4 hours   **d.** 6 hours

61. Switches, panelboards, wireways and transformers are allowed to be mounted above or below one another if which of the following conditions exists:

    a. They don't extend more than 6 inches beyond the front of the equipment.

    b. No piece of equipment is rated over 300 volts.

    c. They are flush along the front edge.

    d. None of the above

62. Aluminum enclosures and fittings are allowed to be used with which of the following:

    a. PVC conduit   **b.** Electrical nonmetallic tubing

    c. Ferrous conduits   **d.** Steel electrical metal tubing

63. Any 125 volt single-phase receptacles must be protected by ground fault circuit interrupters if they are within which of the following distances from the inside wall of a hot tub:

    a. 15   **b.** 13

    c. 12   **d.** None of the above

64. Which of the following is not a file type:

    a. Mill   **b.** Tubular

    c. Bastard   **d.** Half round

65. The neutral conductor in an electrical installation has which of the following qualities:

    a. It carries the unbalanced current.

    b. It is the white conductor.

    c. It does not apply ampacity correction.

    d. All of the above

66. Receptacles must be of the grounding type if they are installed on which of the following:

    a. 40 amp circuit   **b.** 30 amp branch circuit

    c. 15 and 20 amp branch circuits   **d.** None of the above

67. In a corroded electrical connection, high spot temperature is caused by which of the following:

    a. Increase in the voltage drop across the connection

    b. Absence of surge protection

    c. Decrease in the resistance of the connection

    d. Ampacity that is too high for the connection

68. Which of the following symbols represents delta connection:

    a. Ω   **b.** ϖ

    c. △   **d.** ∅

69. A switch is used for which of the following purposes:

    a. Making a connection   **b.** Breaking a connection

    c. Changing a connection   **d.** All of the above

70. Which of the following has the highest electrical resistance:

    a. Water   **b.** Paper

    c. Iron   **d.** Brass

71. Which of the following changes alternating current to DC:

    a. Rectifier   **b.** Capacitor

    c. Transfer switch   **d.** Phase conductors

72. Which of the following is the term for the ability of a material to allow the flow of electrons:

    a. Ampacity   **b.** Resistance

    c. Current   **d.** Conductance

73. Although silver, gold, and copper are all excellent conductors of electricity, copper is the most commonly used for which of the following reasons:

    a. Strength

    b. Higher melting point

    c. Its ability to bond with a wider variety of materials

    d. Lower cost

74. The term "open circuit" describes which of the following conditions:

    a. The circuit is carrying voltage.

    b. All parts of the circuit are not in contact.

    c. There is no disconnecting means applied.

    d. The circuit is experiencing voltage variations or drops.

75. A close nipple is best described as having which of the following characteristics:
   a. Threads inside the cap
   b. No threads
   c. Threads over the entire length
   d. Does not exceed 1/2 inch in diameter

76. A machinery limit switch is used for which of the following purposes:
   a. To close the circuit when the current exceeds a preset limit
   b. To open the circuit when temperature reaches a preset limit
   c. To open the circuit when travel reaches a preset limit
   d. To limit voltage drops

77. Conduit installations should not have which of the following:
   a. Conduits that run uphill
   b. Low points between successive outlets
   c. A high point at an outlet
   d. All of the above

78. Which of the following symbols represents a duplex electrical outlet:

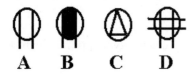

79. A fuse under normal load would most likely become hot because of which of the following conditions:
   a. The rating of the fuse is too low for the application.
   b. Insufficient pressure at the fuse clips
   c. A surge in power has occurred.
   d. A lightning strike

80. One receptacle on a single branch circuit must have which of the following ratings:
   a. 15 amps
   b. 100 percent of the branch circuit rating
   c. 110 volts
   d. None of the above

81. A service disconnect means shall be installed in which of the following locations:

   a. At the nearest point of entrance to a structure

   b. Within 3 feet of the electrical panel

   c. Inside or outside of a building

   d. All of the above

82. Wooden plugs shall not be used to mount electrical equipment to which of the following types of material:

   a. Masonry **b.** Plaster

   c. Concrete **d.** All of the above

83. All wiring must be installed so that when complete the installation:

   a. Is as efficient as possible

   b. Is free of shorts and unintentional grounds

   c. Allows for future expansion of the electrical system or components

   d. All of the above

84. A solenoid is which of the following:

   a. An electromagnet **b.** A relay

   c. A fuse **d.** None of the above

85. Lighting and branch circuit panelboards shall have 10 percent or more of the overcurrent devices rated at 30 amps.

   a. True **b.** False

86. The soft conversion for 3 feet is 914.4 millimeters.

   a. True **b.** False

87. The definition of a device permits it to carry but not utilize electric energy.

   a. True **b.** False

88. A conductor must be sized in accordance with which of the following requirements:

   a. No less than 100 percent of the noncontinuous load

   b. No less than 100 percent of the noncontinuous load, plus 125 percent of the continuous load

   c. No greater than 125 percent of the continuous load

   d. No less than 125 percent of the continuous load

89. A device that establishes an electrical connection to the earth is which of the following:

    a. A lightning rod   **b.** A bonding jumper

    c. Grounded conductor   **d.** Grounding electrode

90. Supplementary overcurrent protective devices used in light fixtures or luminaires are required to be readily accessible.

    a. True   **b.** False

91. A common grounding electrode conductor may connect transformers to a grounding electrode if the common grounding electrode conductor is which of the following:

    a. No smaller than 3/0 AWG copper

    b. Connected by a bonding jumper

    c. Direct-buried no less than 8 feet in the ground

    d. Common grounding electrodes are prohibited from connecting to transformers.

92. A premises wiring system with power derived from a source of electric energy or equipment other than a service is considered to be which of the following:

    a. Low-voltage system   **b.** Solar or photovoltaic system

    c. A separately derived system   **d.** Closed-loop system

93. TVSS devices shall not be installed if the system circuits are in excess of 600 volts.

    a. True   **b.** False

94. Service equipment is grounded to a grounding electrode to keep metal parts, which are subject to a ground fault, at the same potential as the earth.

    a. True   **b.** False

95. Which of the following colors of insulation or markings are not allowed to identify hot phase conductors:

    a. Black and red for 120/208 volt systems

    b. Yellow or orange for 277/480 volt systems

    c. White or gray

    d. All of the above

96. Wiring devices for use on 120 volt circuits are marked as 125 volts.

    a. True   **b.** False

97. In a high-leg delta arrangement on a switchboard with bus bars, phase B would have which of the following:

   a. The highest voltage to ground   **b.** The lowest voltage to ground

   c. The highest ampacity   **d.** The lowest ampacity

98. 200 amp service entrance equipment located indoors at a single-family dwelling must meet which of the following requirements:

   a. Have 24 branch circuits

   b. Use bolt-in fuses or circuit breakers

   c. Be illuminated

   d. None of the above

99. The demand factor for three commercial kitchen loads is which of the following:

   a. 90 percent

   b. 110 percent of the maximum ampacity

   c. Based on the ambient temperature

   d. None of the above

100. The total resistance in resistors connected in series is which of the following:

   a. Equal to the largest resistor in the series

   b. Equal to the sum of all of the individual resistance values

   c. The sum of all of the resistance values divided by the number of resistors in the series

   d. None of the above

ANSWERS

CLOSED BOOK EXAM

1.	B	26.	D	51.	C	76.	C
2.	A	27.	C	52.	C	77.	B
3.	C	28.	A	53.	C	78.	A
4.	D	29.	B	54.	D	79.	B
5.	C	30.	A	55.	B	80.	B
6.	B	31.	D	56.	A	81.	C
7.	A	32.	A	57.	C	82.	D
8.	C	33.	C	58.	B	83.	B
9.	A	34.	C	59.	C	84.	A
10.	D	35.	D	60.	B	85.	B
11.	C	36.	D	61.	A	86.	A
12.	B	37.	C	62.	D	87.	B
13.	B	38.	C	63.	D	88.	B
14.	D	39.	A	64.	B	89.	D
15.	A	40.	D	65.	A	90.	B
16.	D	41.	D	66.	C	91.	A
17.	C	42.	D	67.	A	92.	C
18.	B	43.	D	68.	C	93.	A
19.	D	44.	A	69.	D	94.	B
20.	A	45.	B	70.	B	95.	C
21.	B	46.	C	71.	A	96.	A
22.	A	47.	A	72.	D	97.	A
23.	B	48.	B	73.	D	98.	C
24.	A	49.	B	74.	B	99.	A
25.	D	50.	D	75.	C	100.	B

Chapter 17
TECHNIQUES FOR STUDYING AND TAKING YOUR TEST

STUDYING FOR THE EXAM

- **Establish a Study Schedule and Stick to it:** Do you routinely go to a gym or have a favorite TV show that you watch every night? If so, then you already know how to establish a schedule that includes time for activities that are important to you. Set aside a specific time to study each day and stick to it.

- **Avoid Distractions:** Decide on a quiet place to study. Close the door. Turn off the TV. Don't answer the phone. Studying requires you to be focused so that you can increase your comprehension. Quiet study time will also help to prepare you to take the actual exam which will be in a quiet atmosphere. Believe it or not, many people struggle when they take a test because it is so quiet in the testing facility and they are not accustomed to it.

- **Begin at the Beginning:** Do not assume that because you have electrical experience, you can just give a quick glance to the first few chapters of this book or skip them completely. Each chapter provides you with ideas of what type of subjects are covered in that article of the code, as well as Code Updates and practice questions. Even though your field experience has been based on NEC® standards, deciphering technical terminology in a way that will help you correctly answer complex questions on the code is a process that will be much easier if you go from start to finish, one chapter after the other in order.

- **Studying is more than just reading:** The term "study" means to examine, investigate, analyze, process and look at something carefully. You will find a NOTES section at the end of each of the chapters in this book. Use this page to make a list of material you think that you need to study further, specific questions that you did not understand or struggled with, and key words or phases that you feel like you need to review. Then go back and look at these items more carefully so that you can process the information completely.

- **Easy Does It:** If you find that you are getting really frustrated with some of the subject matter, then take a 5 minute break. Get up, stretch, grab a brain-food snack like a piece of fruit or a handful of nuts, and remind yourself that in the end, your studying is going to payoff for you financially. Stay motivated and take it easy on yourself, but don't end your study session. Go back and review earlier sections of the book. Remember, the codes build up from the basics to the specific and you may just need to reinforce your understanding of some of the previous articles and standards in order to comprehend other regulations.

- **Practice Makes Perfect:** Many of the tests in this book come with instructions to time yourself so that you can get an idea of how long it takes you to answer various types of questions. Your goal is to answer as many of the questions correctly in a fixed period of time and strive to get at least 70% of the answers correct. Do not try to memorize the questions and answers in order to increase your score, because none of these exact questions will appear on your licensing exam. The sample exams are only for you to use to practice reading, interrupting, and selecting the most correct answers during the testing process.

TAKING YOUR EXAM

- **Arrive Early:** The last thing you want to do is get all worked up just before you take your test because you were stuck in traffic on the way to the test site and drove at break neck speeds to get there. Allow yourself enough travel time to arrive at the test site about 30 minutes early. This will also give you extra time for parking, to find the test room, and to check in and provide any identification required the day of the test. If you rush to get to the testing facility, your tendency will be to rush through the test. Go Slow.

- **Read Each Question as it is Written:** Only the read the black print. Don't read into the question, and don't try to second guess what the question is asking. If you have allowed yourself ample time to use this book as a study guide, then you will already know the difference between a pool and a fountain, or a wet location and a damp one. Read each question slowly and don't rush to answer it.

- **Increase Your Odds by Eliminating Answers that are Obviously Incorrect:** The advantage to this practice is that it reduces the number of possible correct answers for you to choose from and your goal is to get as many answers right as you can. If a multiple choice question has four options and you can eliminate one of the possibilities right away then you have already increased your odds of getting the correct answer by 25%.

- **Look for Keys and Clues:** Check the wording of the question for negatives such as "is *not*" and "shall *never*". Look for qualifying terms such as *minimum, maximum, must, may, can, shall, no more than or no less than*. Also, connect the meaning of parts of the question by identifying conjunctions like *and, but* and *or*. Any of these words can change the entire meaning of a question. If you know, for example, that in the field the largest conductor size you can use for a wire run is #12 AWG, you would never tell some one that they *may* use at *least* #12 AWG. Look at the licensing test questions the same way.

- **More Clues than you need:** Some questions will also have more information in them then you need in order to pick the correct answer. Maybe you have run into something similar to this at a restaurant when the meal you want does not match the dinner specials word for word. If you want to eat a medium rare steak, baked potato with butter and sour cream, a salad with blue cheese dressing but no onions, and a small slice of apple pie for dessert, which of the following specials would you have to order? Special #1 which is steak with baked potato or a salad and any desert, or Special #2 which is steak with potato and a salad but no cheesecake? The answer would be Special #2: medium rare **steak,** baked **potato** with butter and sour cream, a **salad** with blue cheese dressing but no onions and a small slice of apple pie (which is **not cheesecake**). You can underline words in the test questions to help you stay focused on the meaning of the questions so that you will not get side-tracked by information that is not pertinent.

- **Come Back Later:** Rather than to waste valuable time on baffling questions, answer all of the questions that you can and then go back and take the time you have left and dedicate it to the questions you did not understand. Often, by the time you have answered other problems, the concept or details that you were struggling with earlier in the test may have been addressed in subsequent questions and the correct answer will now be clear.

- **The MOST Correct Answers:** Many electricians get a question wrong because they chose the first correct answer they come across in the answer set. The licensing exam is scored on the most correct answer, because this indicates comprehensive knowledge of the subjects. For example, if there was a fire in your home which of the following must you do?

    a. Call the fire department
    b. Get out of the house
    c. Run down the road yelling "Fire"
    d. All of the above

It would be easy to just pick "All of the above", but in reality the one thing that you *must* do is "Get out of the house". Think about the exam answers in the same way. Once you have read each question carefully, eliminated answers which are obviously incorrect, and looked for clues, like the word "must", then you will be able to choose the most current answer to each exam question.